Metabolic Regulation
a molecular approach

SOMERVILLE COLLEGE
LIBRARY

Metabolic Regulation
a molecular approach

B.R. MARTIN
BSc, PhD, MA
Department of Biochemistry
University of Cambridge

BLACKWELL SCIENTIFIC PUBLICATIONS
OXFORD LONDON EDINBURGH
BOSTON PALO ALTO MELBOURNE

© 1987 by
Blackwell Scientific Publications
Editorial offices:
Osney Mead, Oxford OX2 OEL
8 John Street, London WC1N 2ES
23 Ainslie Place, Edinburgh EH3 6AJ
52 Beacon Street, Boston,
    Massachusetts 02108, USA
667 Lytton Avenue, Palo Alto,
    California 94301, USA
107 Barry Street, Carlton,
    Victoria 3053, Australia

All rights reserved. No part of this
publication may be reproduced,
stored in a retrieval system, or
transmitted, in any form or by any
means, electronic, mechanical,
photocopying, recording or otherwise
without the prior permission of the
copyright owner

First published 1987

Set by Setrite Ltd
Hong Kong and
printed and bound
in Great Britain

DISTRIBUTORS

USA and Canada
   Blackwell Scientific Publications Inc
   PO Box 50009, Palo Alto
   California 94303

Australia
   Blackwell Scientific Publications
   (Australia) Pty Ltd
   107 Barry Street
   Carlton, Victoria 3053

British Library
Cataloguing in Publication Data

Martin, B.R.
   Metabolic regulation: a molecular
   approach.
   1. Biological
   chemistry   2. Metabolism
   I, Title
   612'.39   QP514.2

ISBN 0-632-01157-2

41864

Contents

Preface, ix

Abbreviations, x

1 The regulation of enzyme activity 1
- 1A Introduction 1
- 1B Rate control in metabolic pathways 1
- 1C Intrinsic and extrinsic control 4
- 1D Identification of rate-controlling steps 6
- 1E Mechanisms of regulation of enzyme activity 12
- 1F Regulation by reversible binding of effectors 13
- 1G The regulation of enzyme activity by covalent modification 21
- 1H Changes in enzyme concentration 24
- 1J Conclusion 27

2 Mechanisms of action of hormones: general introduction 29
- 2A Introduction 29
- 2B Types of mechanism 29
- 2C Relationship of binding to response 31
- 2D Mechanisms for second messenger generation 33

3 Regulation of cyclic AMP concentration by hormones 35
- 3A Introduction: cyclic AMP and the second messenger hypothesis 35
- 3B The adenylate cyclase reaction 36
- 3C Phosphodiesterases 37
- 3D Regulation of adenylate cyclase: the role of guanine nucleotides 38
- 3E Protein components of adenylate cyclase 45
- 3F The mobile receptor hypothesis 47
- 3G The collision coupling model 50
- 3H Inhibition of adenylate cyclase by hormones 53
- 3J Direct inhibition 55
- 3K Other effectors of adenylate cyclase 57

4 Mechanism of action of cyclic AMP 59
- 4A Introduction 59
- 4B Specificity of cyclic AMP-dependent protein kinase 59
- 4C Structure and mechanism of action of the protein kinase 60
- 4D Conclusions 64

5 Ca^{2+} and cellular regulation 66
- 5A Introduction 66
- 5B Control of cellular Ca^{2+} levels 66
- 5C Calcium transport across the plasma membrane 67
- 5D The transport of Ca^{2+} by the endoplasmic reticulum 70
- 5E The sarcoplasmic reticulum 71
- 5F Mitochondrial Ca^{2+} uptake 73
- 5G Ca^{2+} efflux from the mitochondria 76
- 5H Calcium cycling or buffering 78
- 5J The intramitochondrial free Ca^{2+} concentration 80
- 5K The control of cytoplasmic Ca^{2+} levels in the resting cell 80

6 Modulation of cytoplasmic Ca^{2+} by hormones 83

- 6A Introduction 83
- 6B Methods for studying Ca^{2+} concentration changes in cells 84
- 6C The effects of hormones on liver cells 91
- 6D Ca^{2+} as a second messenger in liver: evidence for the involvement of Ca^{2+} rather than cyclic AMP 94
- 6E Effects of Ca^{2+} on liver metabolism 97
- 6F Source of the rise in cytoplasmic free Ca^{2+} 98
- 6G The nature of the intracellular Ca^{2+} pool 102

7 Mechanisms of action of calcium as a regulator 105

- 7A Introduction 105
- 7B Calcium-dependent regulator proteins 105
- 7C Calmodulin 106
- 7D Interaction of calcium/calmodulin complexes with target proteins 109
- 7E Processes affected by calmodulin 112
- 7F Troponin C 116

8 Interactions between cyclic AMP and Ca^{2+} as messengers 117

- 8A Introduction 117
- 8B Levels of interaction 117
- 8C Effects on the metabolism of the other messenger 119
- 8D Interactions at the level of intermediary metabolism 122
- 8E Conclusions 129

9 Hormone action and phosphatidylinositol turnover 130

- 9A Introduction 130
- 9B Cellular location of phosphatidylinositol 130
- 9C Reactions of phosphatidylinositol metabolism 132
- 9D Enzymes of inositol lipid metabolism 133
- 9E Effects of hormones on the metabolism of inositol phospholipids 135
- 9F Role of phosphatidylinositol breakdown in mediating the hormone response 138
- 9G Relationship to calcium mobilization 140
- 9H Possible messenger molecules 140
- 9J Conclusions 142

10 Mechanisms of action of insulin 144

- 10A Introduction 144
- 10B Effects of insulin on cell metabolism 144
- 10C Receptor binding 145
- 10D Negative cooperativity in binding 147
- 10E The insulin receptor 148
- 10F The insulin second messenger 155
- 10G Receptor-mediated production of a second messenger 157
- 10H Mechanisms for alteration of enzyme activity by insulin 162
- 10J Protein kinases involved in insulin action 168
- 10K Conclusions 169

11 Mechanism of action of steroid hormones 170

- 11A Introduction 170
- 11B Structure of steroid hormones 170
- 11C Mechanisms of action 171
- 11D Entry into the cell 172
- 11E Acute effects of steroid hormones 173
- 11F Modification 174
- 11G The steroid hormone receptor 174
- 11H Pure receptor studies 178
- 11J The nature of the nuclear response 182
- 11K Reversal of the action of steroid hormones 183
- 11L Conclusions 185

12 Growth factors 186

- 12A Introduction 186
- 12B Source and structure of EGF and NGF 187
- 12C Physiological effects of growth factors 189
- 12D Mechanism of action of EGF 192
- 12E The nature of the mitogenic signal 195
- 12F Mechanism of action of NGF 201
- 12G Conclusions 208

13 Metabolic integration: general introduction 209

- 13A Introduction 209
- 13B Physiological functions of muscle, liver and adipose tissue 209
- 13C Absorption and transport of metabolites 211
- 13D Interactions between brain, muscle, liver and adipose tissue through the circulation 214

14 Regulation of metabolism in cardiac and skeletal muscle 218

- 14A Introduction 218
- 14B Glycolysis and glycogen metabolism 219
- 14C Regulation of muscle glycolysis 219
- 14D Regulation of glycogen metabolism 227
- 14E Glycogen breakdown 228
- 14F Glycogen synthesis 233
- 14G Protein phosphatases 236
- 14H Glycogen metabolism: conclusions 237
- 14J Control of the citric acid cycle 238
- 14K Conclusions 243

15 Regulation of metabolism in liver 244

- 15A Introduction 244
- 15B Lipid metabolism 245
- 15C The citric acid cycle 251
- 15D Glycogen metabolism in liver 254
- 15E Glycolysis and gluconeogenesis in liver 255
- 15F Conclusions 267

16 Regulation of metabolism in adipose tissue 268

    16A Introduction 268
    16B Fatty acid biosynthesis 269
    16C Short-term regulation of fatty acid biosynthesis 282
    16D Long-term control of fatty acid biosynthesis 286
    16E Lipid uptake in adipose tissue 286
    16F Esterification 287
    16G Lipolysis 288
    16H Conclusions 290

Index 291

Preface

This book is intended for university students of biochemistry and I hope it may also be of interest to people researching in this area. The subject is the regulation of metabolism, with a strong emphasis on the mechanisms by which hormones and other extracellular agonists modulate the behaviour of the cell. Chapters 2–12 are devoted to various aspects of this topic. As well as presenting current concepts of the mechanism of action of hormones, I have tried to give an impression of how ideas develop from the experimental evidence and also of the way in which our picture of the subject has changed over the years.

Where there are conflicting theories I have tried to give a balanced view.

Chapters 1 and 13–16 cover the regulation of enzyme activity and the integration of regulation of metabolism. There are many excellent texts which deal with these subjects in great detail, and the purpose here is to place the mechanism of action of hormones in a wider context rather than to provide a detailed coverage.

I am very grateful to my colleagues who have read parts of the manuscript and made many helpful suggestions. I would like to thank John Moore, Tony Corps, Phillip Tubbs, Stephen Pennington, Martin Brand, Phillip Rubery and especially Aviva Tolkovsky. I would also like to thank Glenda Sharpe for her invaluable help in preparing the manuscript.

Abbreviations

ACTH	adrenocorticotrophic hormone
ADP	adenosine diphosphate
AMP	adenosine monophosphate
ATP	adenosine triphosphate
A23187	a Ca^{2+} ionophore
Ca^{2+}	calcium ion
cAMP	cyclic AMP
Con A	concanavalin A
DNA	deoxyribonucleic acid
EGF	epidermal growth factor
ES	enzyme–substrate complex
FDGF	fibroblast-derived growth factor
FGF	fibroblast growth factor
ΔG	free energy change
$\Delta G°$	standard free energy change
G_i	inhibitory guanine nucleotide-binding protein
Gs	activatory guanine nucleotide-binding protein
GDP	guanosine diphosphate
GTP	guanosine triphosphate
HMG CoA	β-hydroxymethyl glutaryl coenzyme A
5HT	5-hydroxytryptamine
IBMX	isobutyl methyl xanthine
IGF	insulin-like growth factor
K/K_{eq}	equilibrium constant
K_a	apparent dissociation constant
K_i	inhibitor constant
K_m	Michaelis constant
K_s	dissociation constant
mRNA	messenger RNA
Na^+	sodium ion
NAD	nicotinamide adenine dinucleotide
NECA	5'N-ethylcarboxyamide-adenosine
NEM	N-ethyl maleimide
NGF	nerve growth factor
PC12 cells	PC12 phaeochromocytoma cells
PDGF	platelet-derived growth factor
PDH	pyruvate dehydrogenase
PEP	phosphoenolpyruvate
PEP-CK	phosphoenolpyruvate carboxykinase
PFK	phosphofructokinase
Pi	inorganic phosphate
pK	negative logarithm of dissociation constant
p(NH)ppG	guanylyl imidodiphosphate
Q_{10}	temperature coefficient
R	gas constant
s	sedimentation coefficient
T	absolute temperature
TPA	12-0-tetradecanoylphorbol-13-acetate
TSH	thyroid-stimulating hormone
VLDL	very low-density lipoproteins
V_{max}	maximum velocity

Chapter 1 The regulation of enzyme activity

1A Introduction

The regulation of metabolic processes ultimately depends upon the control of enzyme activity. We usually think of enzymes as catalysts for scalar chemical reactions, but many other processes are accelerated by protein catalysts. Carrier proteins are involved in the transfer of molecules across biological membranes, and proteins are also involved in mechanical work such as muscle contraction or the complex structural rearrangements involved in DNA replication. All these types of process are subject to control, but, in relation to the control of metabolism, enzyme-catalysed chemical reactions and carrier-mediated transport are the most important.

1B Rate control in metabolic pathways

Research into mammalian intermediary metabolism has now reached a point where there is a rather complete description of the reactions involved in the metabolic pathways. Our understanding of the catalytic mechanisms and regulation of the enzymes involved in the pathways has also improved over the years. This steady accumulation of information has led to increasingly detailed descriptions of the ways in which flux through metabolic pathways is controlled.

The first description of metabolic control was based upon the idea of a single rate-limiting step for each pathway. In other words, the overall rate of flow in a metabolic pathway is governed by the slowest reaction in the pathway, which may be either a chemical conversion or a transport process across a membrane. The slowest reaction is known as the rate-limiting step. In order to alter the flux through the pathway as a whole, the activity of the enzyme catalysing the rate-limiting reaction must be changed. An alteration in the rate of a reaction which is not rate limiting will have no effect on the overall rate of flux in the pathway. The application of this idea to a straightforward linear pathway is obvious. However, real metabolic pathways tend to be more complicated. Many metabolites may be derived from more than one precursor or give rise to more than one product. In the liver, for example, glucose 6-phosphate is the substrate or product for a total of five different enzymes, each involved in a different metabolic pathway (Fig. 1.1). In this sort of situation the idea of a single rate-limiting step is not adequate, since all the possible pathways need to be controlled in a concerted fashion.

More recent ideas about the control of metabolic pathways recognize that the activity of all the enzymes will have some influence on the rate

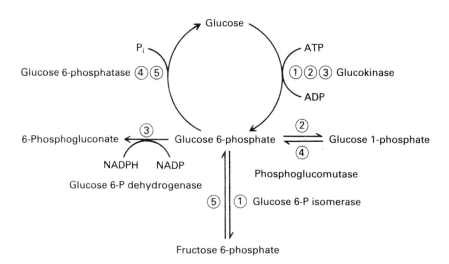

Fig. 1.1. Reactions involving glucose 6-phosphate in mammalian liver. The pathways involved are: (1) glycolysis, (2) glycogen synthesis, (3) pentose phosphate pathway, (4) glycogenolysis, (5) gluconeogenesis.

of the pathway, but that some will be much more important than others. This has been described by Kacser and Burns (1973), in terms of control strength, and by Newsholme, in terms of flux-generating steps. All three interpretations have two things in common. They all stress that the thermodynamics of the reactions in the pathway are important, and, in particular, that reactions catalysed by rate-controlling enzymes are far removed from equilibrium (see Section 1B1). Also, they all agree that, under a given set of circumstances, one enzyme will dominate the regulation of the pathway. However, the more recent interpretations recognize the fact that, should the metabolic situation change, the dominant controlling reaction may also change.

1B1 Thermodynamic considerations

The thermodynamics of individual reactions, and of the pathway as a whole, need to be taken into account when considering regulation. The two parameters of importance are the free energy change, ΔG, and the equilibrium constant, K, which are related to each other by the equation

$$\Delta G = -RT \ln K$$

where R is the gas constant and T the absolute temperature. The standard free energy change, $\Delta G°$, is defined at molar concentrations of substrate and 25°C. Biochemists often use $\Delta G°'$, where the temperature is set at 37°C and the hydrogen ion activity at 10^{-7} mol/l, reflecting the fact that most reactions in intermediary metabolism take place at about pH 7. The other substrate concentrations are still molar. The standard free energy change is useful in that it gives an indication of the favoured direction of a reaction. It relates poorly to the actual free energy change in the cell, since most metabolites are in very dilute solution. Furthermore, the standard free energy change describes the situation occurring as equilibrium is approached and achieved. This is obviously not the situation in the cell where pathways operate at a steady state.

None of the reactions in a pathway are at equilibrium since this would mean no net flux through the pathway. However, many of the reactions will be near to equilibrium; that is, their mass action ratio ([products]/[substrates] in the cell) is close to the equilibrium constant. The actual free energy change of such a reaction in the cell will be small. Changing the activity of the enzyme catalysing such a reaction will have little effect on the rate of the reaction. This is because the mass action ratio is already close to the equilibrium constant. The most that increasing the activity of the enzyme can achieve is to bring the reaction to equilibrium, a small change which will have little influence on other enzymes in the pathway. Such reactions are controlled by their substrate concentration. If the enzyme is able to maintain a mass action ratio close to the equilibrium constant, it must be operating well below its V_{max}. Thus an increase in substrate concentration will lead to an increase in rate (see Section 1H1).

At steady state, a few of the reactions in a metabolic pathway will be far from equilibrium; that is, the mass action ratio is substantially smaller than the equilibrium constant. This implies that the enzyme catalysing the reaction is operating as fast as it can under the particular circumstances, and that this is inadequate to approach equilibrium. In this case an increase in the activity of the enzyme can result in a large change in the mass action ratio, which can potentially influence other enzymes to cause a large change in the rate of the pathway as a whole. All rate-controlling reactions are far away from equilibrium, but it should be stressed that not all reactions which are far from equilibrium are rate controlling.

1B2 The flux-generating reaction

The concept of a flux-generating reaction has been described by Newsholme. Flux-generating reactions have two properties. First, they are far removed from equilibrium at the steady state. Second, at steady state, the enzyme is saturated with substrate; that is, it is operating at V_{max}. This means that an increase in substrate concentration does not affect the activity of the enzyme. A substantial reduction in substrate concentration would be required to reduce the activity of the enzyme (see Section 1H1). The enzyme activity and, accordingly, the flux through the step in the pathway catalysed by the enzyme, are not affected by changes in the substrate concentration. The activity is affected by regulators acting upon the enzyme molecule to increase V_{max}. It is also assumed that only one enzyme at a time in a pathway can act as a flux-generating step. However, if the metabolic circumstances change, the flux-generating step may also change. Many regulatory enzymes do conform to this pattern.

1B3 Control strength

The idea of control strength arose from attempts to quantify the effect of changing the activity of a particular enzyme on the rate of a pathway as a whole. A number of people have written on the subject, but perhaps the best known paper is by Kacser and Burns (1973). The control strength of an enzyme is the change in the steady-state rate of a pathway relative to the change in the enzyme's activity. In other words, if a doubling of the activity of the enzyme leads to a doubling of the rate of the pathway, the enzyme has a control strength of one. It does not matter whether the change in activity is produced by a change in substrate concentration, by a change in K_m, or by a change in V_{max}. The sum of the control strengths of all the enzymes in a pathway is also one. In a simple linear pathway it is usually found that one enzyme has a large control strength approaching one and is the single major site of control. However, in more complex pathways the control is often shared between several enzymes.

Control strength is quite difficult to determine. The usual method is to apply a specific inhibitor of one enzyme in the pathway and determine the effects of the inhibitor on the rate of the pathway as a whole. Before doing this it is necessary to examine the effects of the inhibitor on the isolated purified enzyme, or at least under circumstances where the enzyme has been isolated from effects resulting from changes in the activity of other enzymes near to it in the pathway. A common feature of all these descriptions of metabolic control is that the activity of rate-controlling enzymes is regulated by some sort of chemical signal, which may be a metabolite in the pathway or a molecule whose only function is regulation.

1C Intrinsic and extrinsic control

Metabolism is controlled at two levels. Intrinsic control refers to the control of enzymes by cellular metabolite levels. Its major role is maintaining the ability of the cell to function and, in particular, maintaining the availability of ATP. In a single-celled organism this is the only type of control that operates much of the time.

In a multicellular organism, where the cells are differentiated to form different organs with different physiological functions, a second level of control is required. The metabolism of the cell must respond to the needs of the whole organism. The mammalian liver provides a good example. In the well-fed state, the major pathways are glycogen synthesis, fatty acid synthesis, and pathways involved in the production of energy required for the synthetic reactions. In starvation, the whole direction of metabolism switches to the net production of glucose by glycogenolysis and gluconeogenesis. This type of response is mediated in part by changes in enzyme activity caused by the change in available metabolites—in this case decreased blood glucose and increased free fatty acids—and in part by the action of hormones. Other types of regulation from outside the cell include nervous stimulation and the action of local regulators such as prostaglandins.

In intrinsic regulation, the regulators are mainly metabolites in the relevant pathway. The concept of feedback regulation has been around a long time, but was particularly well illustrated by studies on bacterial amino acid metabolism in the 1950s. In many cases interconversions of amino acids follow a simple linear sequence of reactions, an example being the five reaction pathway which converts threonine to isoleucine (Fig. 1.2). It was found that the first enzyme of the pathway, threonine deaminase, was inhibited by the end-product isoleucine. Clearly this is a very efficient form of control, regulating the flow at the beginning of the pathway when there is enough of the final product. Such negative feedback control is a very common phenomenon. In most cases the pathways are much more complicated but the general principle still holds. The end-product is not necessarily derived directly from the starting material. An example is the generation of ATP by catabolic pathways. In this case the relevant controlling factor is the availability of energy in the cell as ATP.

Other types of control occur, but are much less common. Positive feed-forward control describes the situation where a metabolite early in the pathway activates an enzyme later in the pathway. This allows for concerted regulation of two enzymes. The activation of the first rate-controlling enzyme in the pathway leads to an increase in the concentration of its products (see Section 1D2). One of these products may then activate the second enzyme (Fig. 1.3).

In positive feedback control, a metabolite activates an enzyme earlier in the pathway. In many cases the metabolite is a product of the enzyme

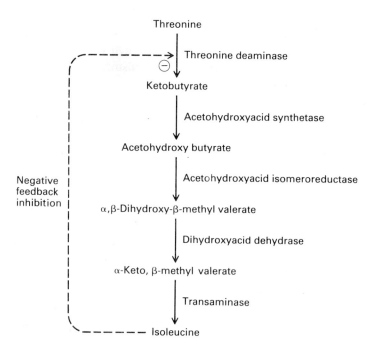

Fig. 1.2. Conversion of threonine to isoleucine in *Escherichia coli*.

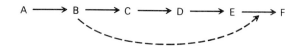

Fig. 1.3. Positive feed-forward control. Activation of the enzyme converting A to B leads to an increase in the concentration of B (see Section 1D2). B then activates the enzyme converting E to F so that the two enzymes are regulated in concentration.

it activates. The concentration of a product will increase upon the activation of a rate-controlling step. If the product is itself an activator of the enzyme, this will potentiate the activation caused by any other effector.

1D Identification of rate-controlling steps

There are a number of factors which help to narrow down the search for the rate-controlling steps in a pathway. First, the enzyme is likely to be near the beginning of the pathway or, in the case of a branched pathway, near to a branch point. Measurement of the maximum catalytic activities of the enzymes in the pathway under optimum conditions may be helpful. Enzymes which have a very high activity relative to other enzymes in the pathway are unlikely to be rate limiting. However, this only gives a preliminary indication since there is often no way of knowing how close to its maximum activity an enzyme is operating in the intact cell.

Rate-controlling enzymes are likely to catalyse reactions with a large negative standard free energy change. However, this is not a totally reliable indication (see Section 1D1).

1D1 Mass action ratios

One of the most satisfactory methods for determining the rate-controlling steps in a pathway is to measure how close the reactions are to equilibrium. To do this, the mass action ratio, [products]/[substrates], is measured in the intact cell. This is done by incubating the tissue under defined conditions and then freezing it very rapidly. The tissue is then extracted, without thawing, by homogenization in a deproteinizing agent such as trichloroacetic acid. The concentrations of metabolites can then be measured and the mass action ratio for each of the reactions in the pathway can be determined (see Section 1D3). If the reaction is close to equilibrium, the mass action ratio will be close to the equilibrium constant. If, however, the reaction is rate controlling in the cell, it will not approach equilibrium and the mass action ratio will be lower than the equilibrium constant. Rate-controlling enzymes usually have a mass action ratio in the intact cell between 100 and 10 000 times lower than the equilibrium constant for the reaction. Such reactions usually, but not invariably, catalyse reactions with large equilibrium constants and large negative standard free energy changes. They are often referred to as irreversible. Of course no chemical reaction is actually irreversible, and the term means that, in the cell, the rate in the reverse direction is negligible.

However, the standard free energy change of the reaction is not the only factor. Some irreversible steps have small standard free energy changes. A good example is the transport of glucose into muscle. $\Delta G°$ is actually zero in this case, since K is one. However, in a functioning cell, the intracellular glucose concentration is several orders of magnitude lower than the extracellular glucose as a consequence of the action of hexokinase. Thus the mass action ratio is much lower than the equilibrium constant, ΔG in the cell is negative, and the process is irreversible. On the other hand, some enzymes catalyse reactions with a large standard free energy change but are still close to equilibrium. This is the case with lactate dehydrogenase which has a very high activity compared with other enzymes which act upon any of its substrates or products. Thus there are two factors which tend to make a reaction irreversible in the cell: a high negative $\Delta G°$, and a low enzyme activity relative to other reactions near to it in the pathway. In most, but not all, rate-controlling reactions, both factors apply.

1D2 Changes in the mass action ratio

The observation that a reaction has a mass action ratio far removed from equilibrium indicates that it may be a rate-controlling step. However, not all reactions which are far from equilibrium are involved in the regulation of metabolic pathways. To be sure that the reaction also regulates the flow in the pathway we need more information, and the system has to be perturbed so that the flux in the pathway is increased or decreased. If the flux in the pathway is increased, it follows that the activity of the enzyme controlling the flux will have increased. This will have the effect of increasing the concentration of its products and reducing the concentration of its substrates so that the mass action ratio moves closer to the equilibrium constant. If the flux through the pathway is reduced as a consequence of the inhibition of a rate-controlling enzyme, the concentration of its substrates will rise and the concentration of its products will fall.

The key observation is that the concentration of the substrates changes in the opposite direction to the rate of the pathway. This is often expressed in the form of a cross-over plot. The concentrations of the intermediates in the pathway are determined with the tissue functioning at steady state, and each concentration assigned the value 100. The pathway is then activated to a new steady state and the concentration

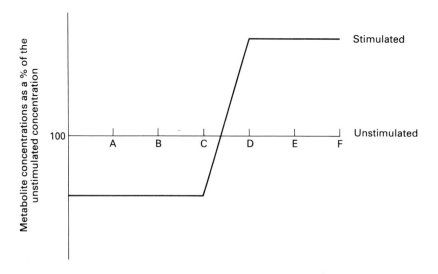

Fig. 1.4. Cross-over plot for isolating a rate-controlling step. On stimulation of the rate from metabolite F, the concentration of metabolites A – C decreases while the concentration of D – F increases. This identifies the reaction from C – D as the rate-controlling step.

of the intermediates determined again. These are then expressed relative to the original concentration and the data are plotted as shown in Fig. 1.4. The point where the lines cross identifies the regulatory step. Thus the identification of a regulatory step in a pathway requires two pieces of information. First, the step must be shown to have a mass action ratio far removed from the equilibrium constant. Second, the mass action ratio must be shown to change if the rate of the pathway changes.

1D3 Problems with the mass action ratio approach

The main problem is the difficulty of accurately measuring the mass action ratio. The usual method is to freeze pieces of whole tissue very rapidly and extract with a strong deproteinizing agent without thawing the tissue. If done carefully there will be little change in the metabolites present. However, this gives the whole tissue content and does not distinguish between the separate contributions of the cytoplasm and intracellular organelles, in particular the mitochondria. For example, pyruvate kinase is a cytoplasmic enzyme which catalyses the reaction:

phosphoenolpyruvate + ADP → pyruvate + ATP

However, ATP, ADP, pyruvate and PEP are all found in the mitochondria as well, where their concentrations will almost certainly differ from those in the cytoplasm. It is technically impossible to break cells and separate mitochondria from cytoplasm rapidly enough to prevent changes in the concentration of rapidly turning over metabolites. Some metabolites, such as pyruvate and oxaloacetate, are produced and utilized very rapidly in relation to their concentration in the cell. Under some circumstances they may turnover several times a second. Another possible difficulty arises from chemical compartmentation. Many different enzymes bind substrates such as NAD^+ and ATP. Thus the total substrate concentration measured in tissue extracts may be considerably greater than the free concentration available in the cell.

In practice, the mass action ratio of a rate-controlling step is usually so far removed from equilibrium, and changes so much on activation of the pathway, that these problems can be ignored. However, the control strength theory suggests that enzymes can exert significant regulation on a pathway when the ratio of the equilibrium constant to the mass action ratio is as low as 10. Under these circumstances the difficulty of accurate measurement of the mass action ratio would be important. The alternative approach to determining the position of rate-controlling steps in a pathway is the use of inhibitors (see Section 1B3).

1D4 Determination of the rate-controlling steps in heart muscle glycolysis

In this section the determination of the rate-controlling steps in heart muscle glycolysis will be described to illustrate the application of the general approach discussed above. Figure 1.5 shows the pathway and Table 1.1 lists the enzymes in sequence together with their maximum activities, equilibrium constants, and the mass action ratios in aerobic conditions. The maximum activities of hexokinase, phosphofructokinase, aldolase and enolase are considerably lower than those of the other enzymes in the pathway, so they must be considered as candidates for the control point. However, a different picture emerges when the mass action ratios are considered. Hexokinase and phosphofructokinase have mass action ratios more than five orders of magnitude less than the equilibrium constant. In the case of aldolase, the mass action ratio is 10% of the equilibrium constant, and in the case of enolase, 25–50%. Of the four enzymes above, hexokinase and phosphofructokinase are likely candidates for rate-controlling enzymes, and aldolase is possible.

Table 1.1. Glycolytic enzymes in heart muscle

Enzyme	V_{max} (µmol/g/min)	K_{eq}	Mass action ratio
Hexokinase	7	4000	0.08
Phosphoglucoisomerase	65	0.4	0.24
Phosphofructokinase	14	1000	0.03
Aldolase	24	0.0001M	0.00001M
Triose phosphate isomerase	580	0.04	0.24
Glyceraldehyde 3-P dehydrogenase/ phosphoglycerate kinase	135/74	$200M^{-1}$	$9M^{-1}$
Phosphyglycerate mutase	27	0.2	0.12
Enolase	15	3.0	1.4
Pyruvate kinase	145	10 000	40

11 The regulation of enzyme activity

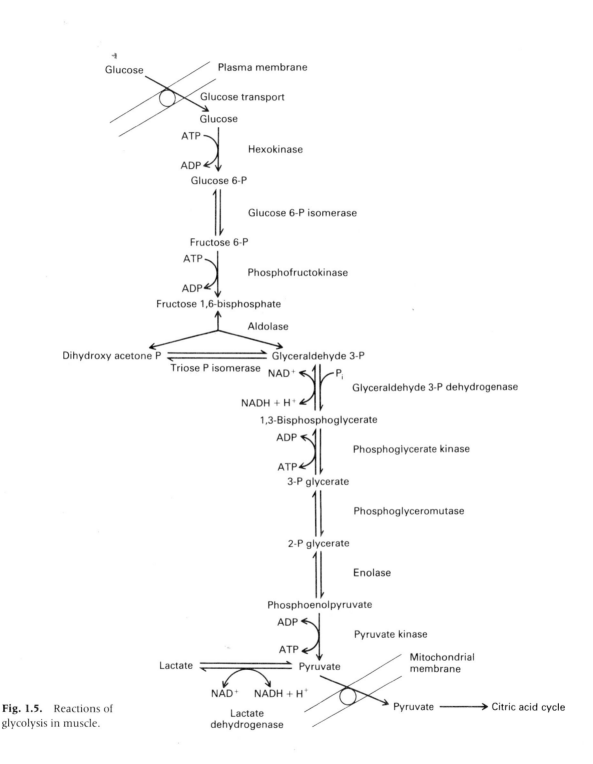

Fig. 1.5. Reactions of glycolysis in muscle.

Pyruvate kinase also has a mass action ratio very far removed from equilibrium and must therefore be considered as a possible control site.

The next question to consider is whether any of these mass action ratios increase when the rate of glycolysis is increased by making the heart anaerobic. In the case of hexokinase, the substrates are glucose and ATP, and the products are ADP and glucose 6-phosphate. It is necessary to perfuse the heart in the presence of insulin for the concentration of glucose in the tissue to be detectable. If this is done, it can be shown that the glucose concentration falls drastically when the heart is made anaerobic. The ATP concentration falls and the ADP and glucose 6-phosphate concentrations rise. Similarly, in the case of phosphofructokinase, the ATP and fructose 6-phosphate concentrations fall, while ADP and fructose 1,6-bisphosphate increase when the heart becomes anoxic. Both enzymes fulfil all the criteria for regulatory enzymes. Pyruvate kinase does not appear to be regulated in spite of the fact that it looks as if it might be a rate-controlling step. When glycolysis is stimulated by anoxia, the concentration of the substrate phosphoenolpyruvate does not fall and the concentration of at least one of the products, ADP, increases rather than decreases.

There is one further control site: the entry of glucose into the cell. As discussed above, the transport process is far from equilibrium and is therefore a potential rate-controlling step. Insulin will increase the intracellular glucose concentration while the extracellular glucose concentration, if anything, falls. Thus glucose transport is regulated by insulin.

1E Mechanisms of regulation of enzyme activity

Having identified the enzyme, or enzymes, which are responsible for the regulation of the rate of a metabolic pathway, the next step is to determine the mechanisms by which the enzymes themselves are regulated.

The activity of a regulatory enzyme changes as a consequence of an alteration in the concentration of one or more signal molecules. In the case of intrinsic control, the signal will usually be a metabolite. In the case of extrinsic regulation, the signal will be produced as a consequence of the extrinsic regulator interacting with the cell and may, or may not, be a metabolite in the pathway. In either case, an efficient control mechanism requires that a small change in the concentration of the signal molecule should result in a large change in the rate of the pathway. In some cases, such as the activation of glycolysis in insect flight muscle, the increase in rate may be as much as one hundredfold.

In other words, the response needs to be amplified relative to the signal. There are a variety of mechanisms by which such amplification is achieved.

There are three general mechanisms by which the activity of enzymes can be regulated: control by reversible binding of effectors, by covalent modification, and by alteration of enzyme concentration.

In the first case, the enzyme is activated or inhibited as a direct consequence of the binding of a signal molecule, usually but not always a metabolite, which may, or may not, be a substrate or product of the enzyme reaction. This type of control gives a very rapid, essentially instantaneous response. However, the extent to which the activity of an enzyme can be changed by this type of mechanism is limited.

Covalent modification of an enzyme requires a second enzyme which catalyses a covalent change in the first enzyme and is itself subject to control. The modifier enzyme may be regulated by either intrinsic or extrinsic signals. In this case the first enzyme which takes part in the metabolic pathway exists in two different structural states with different activities. The response to this type of control will be slightly slower, but in most cases it is complete within minutes.

Alterations in the concentration of enzymes in the cell result from changes in the relative rates of synthesis and degradation of the enzyme. The potential for alteration in activity is very large; during cellular differentiation, for example, enzyme activities appear which were previously absent altogether. This type of response is long term, usually taking hours to be complete in mammals. The intervention of extrinsic regulators is usually required. In the following three sections the three different types of mechanism for the regulation of enzyme activity will be considered.

1F Regulation by reversible binding of effectors

The regulation of enzymes by reversible binding of effectors shows an enormous range of complexity. In some cases the activity of the enzyme may simply be regulated by the availability of the substrate. In other cases seven or eight different effectors may change the activity of the enzyme.

1F1 Substrate control

The rate of all enzyme-catalysed reactions is affected by the availability of substrate. This was first described as early as 1904 by Michaelis and Menten who showed that the rate of many enzyme-catalysed reactions

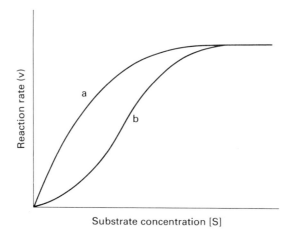

Fig. 1.6. Non-cooperative and positively cooperative kinetics. (a) Non-cooperative kinetics. (b) Positive cooperativity.

varied with the concentration of substrate in a hyperbolic fashion (Fig. 1.6a). They assumed that the factor which limited the rate of the reaction was the binding of substrate to the enzyme, and that the binding reaction is essentially at equilibrium. From this it follows that the rate of the reaction will be determined by the concentration of the enzyme containing bound substrate. Thus half the maximum possible rate of reaction will be achieved when half the enzyme contains bound substrate. The substrate concentration needed to achieve this was termed the Michaelis constant, K_m, and in this initial interpretation is equal to the dissociation constant, K_s, of the enzyme − substrate complex. Later it was realised that the process was more complicated and that later stages in the reaction may need to be taken into account. This led to a steady state interpretation rather than the earlier equilibrium interpretation, but the basic hyperbolic relationship between substrate concentration and rate of reaction still holds. The relationship is described by the equation:

$$v = \frac{V_{max} [S]}{K_m + [S]}$$

The derivation of the equation is given in Fig. 1.7.

If an enzyme shows Michaelis−Menten kinetics, the potential for controlling the rate of reaction by changes in substrate concentration is limited. To a rough approximation at concentrations below the K_m, an increase in substrate concentration will result in a proportional increase in rate; that is, doubling the substrate concentration will double the rate. This is the most sensitive response possible. K_m is equal to the substrate concentration which gives half maximum velocity, and a doubling of

A single substrate enzyme reaction can be described by the equation

$$E + S \underset{k_{-1}}{\overset{k_1}{\rightleftharpoons}} ES \underset{k_{-2}}{\overset{k_2}{\rightleftharpoons}} E + P$$

where k_1, k_{-1}, k_2 and k_{-2} are rate constants.

The aim is to relate the initial velocity v of the catalysed reaction to enzyme and substrate concentrations and to the rate constants for the individual steps.

Assumptions
(1) Since initial velocities are measured, the back-reaction can be ignored (i.e. [P] = 0).
(2) Initial concentration of enzyme is very much less than initial substrate concentration.
(3) The concentration of ES complex remains essentially constant, because its rate of formation is balanced by its rate of breakdown (i.e. a steady state is assumed).

The rate of appearance of product is given by $v = k_2 [ES]$. (1)

The rate of formation of $ES = k_1 [E] [S]$. (2)

The rate of breakdown of $ES = (k_{-1} + k_2) [ES]$. (3)

In the steady state, d[ES]/dt = 0, so, from (2) and (3),

$[ES] = k_1 [E] [S]/(k_{-1} + k_2) = [E] [S]/(k_{-1} + k_2)/k_1$ (4)

Combining the three constants in (4), we can write

$[ES] = [E] [S]/K_m$ (5)

where $K_m = (k_{-1} + k_2)/k_1$ (the Michaelis constant).

The concentration of free enzyme [E] is given by

$[E] = [E_T] - [ES]$ (6)

Substituting for [E] in (5) we obtain

$[ES] = \{([E_T] - [ES]) [S]\}/K_m$ (7)

Solving for [ES],

$[ES] = [E_T] [S]/(K_m + [S])$ (8)

From (1) and (8),

$v = k_3 [E_T] [S]/(K_m + [S])$ (9)

When $[E_T] = [ES]$, $v = V_{max}$, so that

$v = V_{max} [S]/(K_m + [S])$ The Michaelis–Menten equation. (10)

Fig. 1.7. Derivation of the Michaelis–Menten equation.

rate to approach V_{max} will require an approximately tenfold increase in substrate concentration above K_m. Thus, in this range, very large changes in substrate concentration are required to produce significant changes in reaction rate. If we consider what happens over most of the activity range, to increase the rate from 0.1 V_{max} to 0.9 V_{max} requires an eighty-onefold increase in substrate concentration. An efficient regulatory process requires a large change in reaction rate in response to a small change in the concentration of the signal molecule, so, as we might expect, changes in substrate concentration rarely act as signals for the regulation of rate-controlling enzymes. A notable exception is glucokinase in liver (see Chapter 15). Changes in substrate concentration are important in determining the rate of reactions in a pathway which are close to equilibrium. This type of reaction never operates close to V_{max}, and the rate will be affected by changes in substrate concentration resulting from the regulation of rate-controlling enzymes in the pathway.

1F2 Cooperative effects

Many enzymes show a sigmoid relationship between the concentration of the substrate and the rate of reaction (Fig. 1.6b). Over the steep part of the curve, a large change in rate results from a relatively small change in substrate concentration. This increases the potential for regulating the enzyme. The first process to be analysed in detail was the binding of oxygen to haemoglobin. In this case there is no subsequent chemical change and only the binding reaction need be considered. The sigmoid response can be explained if it is assumed that the enzyme has more than one binding site for the substrate, and that the binding of one molecule of substrate to one site facilitates the subsequent binding of other molecules of substrate to other sites. If binding of substrate to one site reduces the affinity of the other sites for substrate, then the result will be negative cooperativity. There are few convincing instances of negative cooperativity in the regulation of enzyme activity, but positive cooperative effects are very common. Enzymes which show cooperative effects usually have several subunits with a single substrate binding site on each subunit. However, there are a few cases of single subunit enzymes which show cooperative effects.

A number of models have been proposed to explain the phenomenon of cooperativity. One of the first, proposed by Jacob, Monod, Wyman and Changeux (1963), is often called the concerted model (Fig. 1.8). This suggests that the enzyme can exist in two states: relaxed (R) with a high affinity for the substrate, and tight (T) with a low affinity for the substrate. The two states of the enzyme are in equilibrium, and the

17 *The regulation of enzyme activity*

assumption is that in the absence of substrate, the tight form predominates. The binding of a single molecule of substrate to a subunit in the relaxed state stabilizes the enzyme in this state. This affects the equilibrium in favour of the (R) state so that the number of sites which are receptive to substrate increases. Thus the binding of one molecule of substrate increases the likelihood of binding of further molecules of substrate. A key feature of this model is that the binding of the substrate simply stabilizes the relaxed state and therefore displaces the equilibrium in its favour. Substrate binding does not actually promote the change from the less active to the more active state.

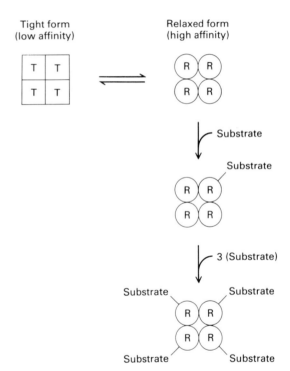

Fig. 1.8. Concerted model for positive cooperativity. Binding of the first substrate molecule perturbs the equilibrium in favour of the relaxed R form thereby facilitating the binding of three further molecules of substrate.

A second model was proposed by Koshland, Nemethy and Filmer (1966), and is often referred to as the sequential model. It followed from the idea of induced fit which suggests that the binding of substrate to an enzyme can induce a conformational change in the enzyme protein; that is, the binding of the substrate helps to alter the shape of the enzyme to obtain the correct catalytic conformation. In a multisubunit protein, a change in conformation of one subunit may result in the change in conformation of adjoining subunits, thereby improving the fit

18 *Chapter 1*

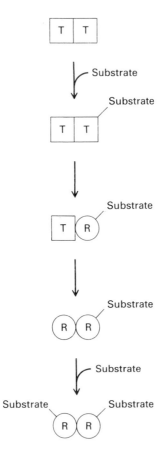

Fig. 1.9. Sequential model for positive cooperativity. The binding of the first substrate molecule causes a conformation change from the T low-affinity form to the R high-affinity form. This then causes the second subunit to change from the T form to the R form, facilitating the binding of the second molecule of substrate.

for the substrate and increasing the probability of substrate binding (Fig. 1.9). In this model the binding of the substrate actively promotes the change from the less active to the more active state.

Both of these models describe kinetic behaviour on the basis of the assumption that the rate of reaction is proportional to the amount of enzyme containing bound substrate. Where the function of the protein is essentially binding, as in haemoglobin, such an approach is obviously valid. However, in the case of an enzyme-catalysed reaction, a steady state approach is theoretically more valid. Another problem with the two models above is that they are unable to explain the existence of cooperative effects in enzymes which have only one binding site for the substrate. Such enzymes are uncommon, but do occur. There are many steady state models for cooperative effects in enzymes, some of them very complex. One of the simplest is the model proposed by Rabin (1967).

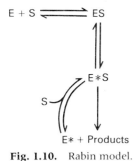

Fig. 1.10. Rabin model.

The Rabin model is summarized in Fig. 1.10. After formation of the enzyme–substrate complex, ES, the enzyme undergoes a conformation change, represented as the formation of E*S, and then yields E* and the products. The model proposes that the slowest step in this process is the formation of E*S from ES. Furthermore, E* can react with substrate directly to form E*S, which bypasses the slowest step in the reaction leading to an activation. The relative rates of formation of E*S by binding of S to E*, and of formation of E from E*, will depend upon the substrate concentration. A higher substrate concentration will tend to maintain the enzyme in the more efficient E* form, so there will be positive cooperativity with respect to substrate concentration. A simple way of looking at this is to say that E* represents a state which 'remembers' its previous encounters with the substrate. This model will fit equally well for single subunit or multisubunit enzymes.

1F3 Allosteric effects

If substrate binding can alter the conformation of an enzyme, then it is obvious that the binding of other non-substrate molecules may also change the enzyme conformation. This leads to the idea of allosteric control in which a molecule, which is not the substrate, binds to a specific regulatory site on the enzyme, producing a conformational change which may either activate or inhibit the enzyme.

Allosteric effects may be explained by either the concerted model or the sequential model. In the case of the concerted model the binding of the effector will stabilize either the relaxed state, leading to activation, or the tight state, leading to inhibition. In the case of the sequential model, the effector will promote a change to a more or less effective form of the enzyme.

In the case of an effector which is not a substrate of the enzyme, there is no obvious requirement for more than one binding site for substrate, and hence, multiple subunits. An example of a common allosteric enzyme which has only one subunit, is hexokinase. However, most allosteric enzymes have multiple subunits and interaction between the subunits is necessary, both for cooperative effects involving only the substrate, and for effects involving effectors other than the substrate. In such enzymes, the disruption of subunit/subunit interactions can lead to a loss of all the regulatory responses of the enzymes without the loss of the catalytic activity. The involvement of effectors other than the substrate increases the possibilities for regulation enormously. It is obviously necessary for feedback control. It also allows for a much larger regulatory response. Often the cell requires a key regulatory enzyme to respond to a rather general alteration in the metabolic state, for example a change from low to high blood glucose. Such a change will result in changes in the concentrations of many metabolites, all of which are potential regulators of the enzyme. If several different metabolites do alter the activity of the enzyme, then acting together their total effect may be large, even though the effect of individual metabolites on the activity of the enzyme may be modest. An obvious example is phosphofructokinase which responds to a large number of effectors which all change in concentration when the demand of the cell for ATP changes (see Chapter 14).

1F4 Substrate cycles

A substrate cycle consists of two or more enzymes which catalyse opposing reactions in a metabolic pathway. An example is the substrate cycle between phosphofructokinase and fructose 1,6-bisphosphatase (Fig. 1.11). Both reactions have a high negative free energy change and are irreversible. Such cycles are sometimes referred to as futile cycles since

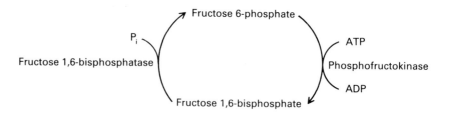

Fig. 1.11. The fructose 6-phosphate/fructose 1,6-bisphosphate substrate cycle.

the possibility exists for ATP hydrolysis in the absence of net flux in the pathway. However, such cycles have two important roles in the cell. First, they may allow the carbon flow in the pathway to go in either direction, thus a liver cell may either break down glucose by glycolysis or synthesize glucose by gluconeogenesis. The second function is as a means of amplifying regulatory responses. In muscle, phosphofructokinase is activated by AMP, while fructose 1,6-bisphosphatase is inhibited. Suppose in a resting muscle the rate of flow from fructose 6-phosphate to fructose 1,6-bisphosphate through phosphofructokinase is two, while the rate in the reverse direction through fructose 1,6-bisphosphatase is one. The net flow in the pathway will be one. If the AMP concentration changes so that phosphofructokinase is activated twofold and fructose 1,6-bisphosphatase activity halved, the net flow will increase from 1 to 3.5. Thus a twofold change in activity of each of the two enzymes results in a 3.5-fold change in flux—a considerable amplification. If the flow through phosphofructokinase was 100, relative to a flow through fructose 1,6-bisphosphatase of 99, a twofold change in activity of both enzymes would cause a hundred and fiftyfold change in net flux. Thus the more energy committed to cycling between the substrates, the greater the amplification of the regulatory response. There is evidence that in man, the rate of cycling in skeletal muscle is under hormonal control, and that the stress hormones, such as adrenaline, increase the rate of cycling and hence the sensitivity of glycolysis to activation.

1G The regulation of enzyme activity by covalent modification

The structure of enzymes and other proteins can be altered by the action of other enzymes. Two types of modification can occur. In the first, the enzyme is subjected to a highly specific proteolytic cleavage. Usually this involves the splitting of a single peptide bond which may, or may not, result in the release of part of the protein from the original native protein. This type of reaction is irreversible. The second type of covalent modification involves the incorporation of a chemical group into the protein. There are a number of different examples. Phosphate may be incorporated using ATP as a donor, ADP-ribose may be incorporated using NAD as a donor, AMP may be incorporated using ATP as a donor, or the enzyme may be methylated. This type of covalent modification can be reversed by the action of a second enzyme which removes the group incorporated by the first.

In mammals, by far the most important mechanism involved in the regulation of intracellular enzymes is phosphorylation by protein kinases. Proteolytic cleavage is common as a means of activating extracellular proteins. For example, the digestive proteases are secreted as inactive enzymes and activated in the gut by limited proteolysis. The activity of intracellular enzymes is often affected by limited proteolysis, and in many cases the enzymes affected are the same as those which are regulated by changes in their phosphorylation state. As yet however, there is no clear evidence that such effects are physiologically important.

1G1 Regulation by phosphorylation

Phosphate is incorporated into the enzyme by a protein kinase using ATP, and can be removed by a protein phosphatase (Fig. 1.12). In most cases the phosphate forms an ester with the hydroxyl group on the side chain of a serine residue, but there are instances where threonine, or

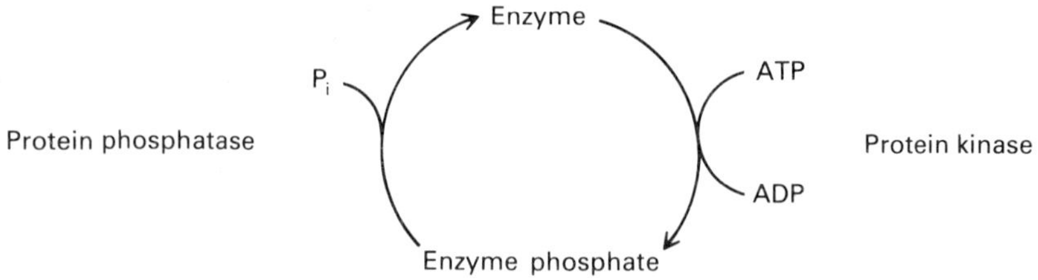

Fig. 1.12. Protein phosphorylation.

tyrosine, hydroxyl groups are phoshorylated (Fig. 1.13). Recently there has been considerable interest in protein kinases which phosphorylate basic amino acid residues. However, the physiological importance of this is not clear. The first protein kinase to be described was the cyclic AMP-activated protein kinase which is described in detail in Chapter 4. This enzyme responds to alterations in the level of cyclic AMP which is under hormonal control (see Chapter 3). It provides a major means of extrinsic control of enzyme activity. More recently, many protein kinases have been described which respond to a variety of different effectors, and it appears that protein phosphorylation/dephosphorylation is a very widespread type of regulation. As a mechanism it has the potential for a

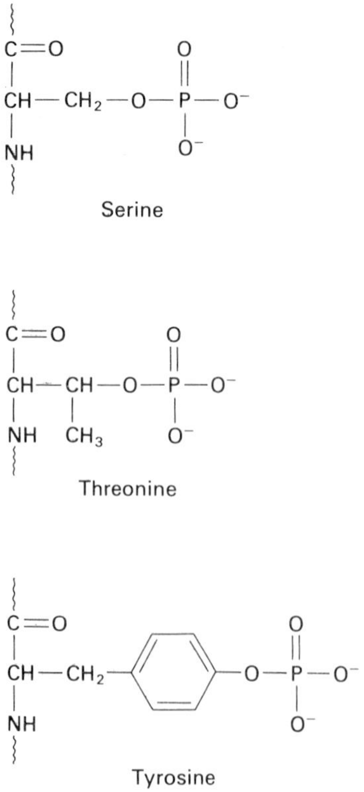

Fig. 1.13. Phosphorylation sites in proteins. Phosphorylated amino acids are shown as components of a peptide chain.

large amplification of the signal, that is, a large change in enzyme activity in response to a small change in the signal molecule concentration. It also allows many different effectors to operate upon a single system.

1G2 Nature of the change in activity after phosphorylation

In some cases phosphorylation changes the enzyme between a completely inactive state, and a state which has activity and may also be modulated by other effectors. This is the case with pyruvate dehydrogenase where the phosphorylated state has no activity at all, and the dephosphorylated state is inhibited by the products of the reaction.

More often both states of the enzyme have some activity, and activation results, at least in part, from changes in sensitivity to other effectors. In the case of liver pyruvate kinase, phosphorylation increases the K_m for phosphoenolpyruvate, the substrate, increases the K_a for the activator fructose 1,6-bisphosphate, and reduces the K_1 for both inhibitors, alanine and ATP. All these effects tend to reduce the rate of the reaction.

1G3 Amplification

Amplification of the signal arises from the possibility of a cascade of enzymes, a good example being the control of glycogenolysis in muscle (Fig. 1.14). In this case a hormone binds to its receptor and activates

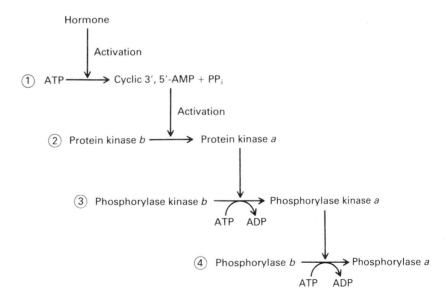

Fig. 1.14. The glycogenolytic cascade. Amplification occurs at steps (1), (3) and (4) (see text). The different stages in the cascade are dealt with in detail in Chapters 3, 4 and 14.

adenylate cyclase, leading to the synthesis of many molecules of cyclic AMP. Each of these may bind to, and activate, a molecule of protein kinase. Each protein kinase may phosphorylate and activate many molecules of phosphorylase kinase, and each activated phosphorylase kinase may phosphorylate and activate many molecules of phosphorylase, the enzyme responsible for the final metabolic response. The potential amplification of the initial signal, the change in hormone concentration, is huge. The control of glycogen metabolism is considered in detail in Chapter 14. Activation cascades can equally well function where the covalent modification is proteolytic, the obvious example being blood clotting.

1G4 Response to multiple effectors

If an enzyme is controlled by its phosphorylation state, an effector may work through three different possible mechanisms. It may bind to the kinase and modulate its activity, it may bind to the phosphatase and modulate its activity, or it may regulate the enzyme itself directly. It may, of course, do all three, thus increasing the response. An effector can also affect the phosphorylation state of an enzyme by altering its conformation, so that it becomes a better or worse substrate for the protein kinase or phosphatase. For example, the rate of phosphorylation can be increased either by an effector activating the protein kinase, or by an effector altering the conformation of the substrate protein so that it is more easily phosphorylated by the protein kinase. This allows for very subtle interactions between extrinsic regulatory factors acting from outside the cell and intrinsic factors which change in response to the metabolic state of the cell.

1H Changes in enzyme concentration

The third mechanism which regulates enzyme activity is the alteration of the concentration of enzyme protein in the cell. This is a relatively long-term control which usually takes several hours to complete. Accordingly, changes in enzyme concentration take place in response to relatively long-term changes in the physiological state. The change may be environmental, as in the onset of starvation, or a change from a carbohydrate-rich to a fat-rich diet. Developmental changes also cause large alterations in enzyme concentration.

1H1 Synthesis and degradation

The concentration of a protein in the cell is governed by the rate of synthesis and the rate of degradation. At steady state the two rates will be equal. The concentration of an enzyme can be increased by either increasing the rate of synthesis or decreasing the rate of degradation, or both at the same time. The stimulation of synthesis of specific proteins by steroid hormones is relatively well understood and is described in Chapter 11. In this chapter, the regulation of enzyme turnover and degradation will be described. This is relatively poorly understood.

1H2 Rates of turnover of enzymes

The rate of turnover of an enzyme is conveniently expressed in terms of its half-life, the time taken for half the protein present to be degraded. Rates of enzyme turnover have been extensively studied in liver where the half-life has been found to vary from about twelve minutes to up to three weeks. The average half-life for protein in the liver is about two to three days. Liver enzymes with a short half-life, reflecting a rapid rate of turnover, are listed in Table 1.2. Many of them are important in regulation, and several can be shown to be rate controlling for their pathway. In all cases the concentration changes in response to a changed physiological state and many of the enzymes are subject to regulation by other types of mechanism. In contrast, enzymes which have a slow rate of turnover have no regulatory significance.

A high rate of turnover allows for a rapid change in the enzyme concentration, if either the rate of synthesis or the rate of degradation is altered. To some extent this is analogous to a substrate cycle where a rapid rate of cycling allows for a large regulatory response.

Table 1.2. Some rapidly turning over enzymes in liver

	Half-life (h)
Ornithine decarboxylase	0.2
RNA—polymerase I	1.3
Tyrosine aminotransferase	2.0
Tryptophan oxygenase	2.5
Deoxythymidine kinase	2.6
β-Hydroxymethylglutaryl-CoA reductase	3.0
Serine dehydratase	4.0
Phosphoenolpyruvate carboxykinase	5.0
Glucokinase	12.0

1H3 Reasons for high turnover rates

The rate of synthesis of a particular protein may be controlled at several different levels. The rate of transcription of the gene may be controlled. Other possible sites of control are the processing of the transcript to give mRNA, the transfer of mRNA out of the nucleus, the rate of degradation of mRNA in the cytoplasm, or the rate of translation of mRNA to make the protein on the ribosome. There is very strong evidence that the rate of transcription is under rigorous control, and that this control is important in determining the enzyme profile of a particular cell type. A comparison of the rates of gene transcription between brain and liver reveals marked differences. Furthermore, the mechanisms by which steroid hormones activate the synthesis of particular proteins is fairly well characterized (see Chapter 11), and is known to involve activation at the level of transcription. In lower organisms there is evidence for control at the level of translation, and recently it has been suggested that translational control occurs in mammals as well. However, this is not, as yet, well established.

1H4 Regulation of degradation

The rate of degradation of a protein is obviously equally important in determining its rate of turnover. Obviously a high rate of turnover demands a rapid rate of degradation as well as a rapid rate of synthesis. The mechanisms involved in the degradation of intracellular proteins are relatively poorly described. A major source of degradative activity is found in the lysosomes. These contain acid proteases which function at the intralysosomal pH of about 4. The lysosomes seem to account for most of the protein breakdown, but there are also cytoplasmic neutral proteases which have been reported to be regulated by a variety of factors including calcium. However, there is no direct evidence for the activity of any individual protease upon a particular enzyme substrate being regulated. A number of theories have been proposed to account for the regulation of the rate of degradation of specific enzymes. However, there is little hard experimental evidence. This probably reflects the difficulty of measuring degradation rates for individual enzymes as opposed to measuring the rate of protein degradation as a whole.

One suggestion is that high rates of degradation result from the properties of the substrate enzymes rather than from the effects upon the proteases. Enzymes with a high rate of turnover tend to be generally rather physically unstable. Large proteins with complex structures tend to be more unstable than small proteins, and to turnover faster. Most enzymes of regulatory importance are large and complex. A simple hypothesis would suggest that the spontaneous denaturation of an enzyme molecule, as a result of its inherent instability, is the first step in breakdown and the main factor in rendering a protein susceptible to proteolysis.

It has also been suggested that the interaction of effectors with an enzyme can affect its rate of degradation. In many cases an increased cellular content of the substrate can be shown to slow the rate of breakdown of an enzyme and increase the half-life. This lends support to the stability hypothesis since substrates almost always improve the physical stability of their enzymes.

More interesting effects have been observed with pathway end-products. Increased concentration of the end-product of the pathway in some cases correlates with an increased degradation of a regulatory enzyme early in the pathway. This suggests that end-product inhibitors may in some cases also have a destabilizing effect on enzymes.

It is not clear whether hormones alter the rate of breakdown of individual enzymes as well as altering their rate of synthesis. Hormones which have a general growth promoting or developmental effect necessarily increase the overall net rate of protein synthesis. During the onset of such a response there is a general inhibition of protein degradation which will help to increase the protein content. This effect will be particularly pronounced in the case of enzymes with a high rate of turnover which may be subject to specific activation of their synthesis over and above the general stimulation. However, certain enzymes decrease in concentration against a general increase in protein content, as is the case with the gluconeogenic enzyme phosphoenolpyruvate carboxykinase, in the presence of insulin in liver. Conversely, in starvation, key gluconeogenic enzymes increase in concentration against a general reduction in protein content. This could be achieved simply by modulating the rate of synthesis. However, the rate of degradation could change in response to hormones as a consequence of changes in the concentration of effector metabolites, or indeed, as a consequence of covalent modification in response to the hormone. There is, however, no direct evidence for this happening.

1J Conclusion

The object of this chapter is to provide a brief survey of the mechanisms available for the regulation of enzyme activity. Many publications give a much more detailed treatment of this subject and the aim here is to provide a background for the chapters on hormone action which follow.

Suggestions for further reading

Cohen P. (1983) *Control of enzyme activity.* Chapman and Hall.
Engel P.C. (1981) *Enzyme kinetics: The steady state approach.* Chapman and Hall.
Goldberg A.L. & St. John A.C. (1976) Intracellular protein degradation in mammalian and bacterial cells. *Annu Rev Biochem,* **45**, 747–803.
Groen A.K., Meer R. van der., Westerhoff H.V., Wanders R.J.A., Akerboom T.P.M. & Tager J.M. (1982) Control of metabolic fluxes. In *Metabolic Compartmentation* (Ed. by H. Siess). Academic Press.
Newsholme E.A. & Leech A.R. (1983) *Biochemistry for the Medical Sciences,* Chapters 2 and 3. Wiley.

Chapter 2 Mechanisms of action of hormones: General introduction

2A Introduction

A hormone may be defined as a molecule which is released into the circulation and specifically modifies the metabolism of one or more types of cell, usually referred to as the target cells. Beyond this common definition, hormones are very diverse both in their chemical structure and in their mechanisms of action. Their structures range from small molecules such as catecholamines, to large polypeptides. The metabolic responses vary widely depending upon the particular hormone and the particular target cell. A hormone may cause quite different metabolic responses in different target cells. This may be a reflection of the metabolism of the target cell, for example the regulation of lipolysis must of necessity be confined to cells which contain stored triacylglycerol. In some cases, however, different responses of different cell types reflect the fact that a single hormone may have more than one mechanism of action resulting from binding to different types of receptor. Catecholamines have at least three different mechanisms of action mediated by different classes of receptor.

The time scale over which hormone effects become apparent also varies very widely from a few seconds to several hours. Changes in membrane transport processes can be detected within seconds, while changes in rates of protein synthesis or cell growth and proliferation may take several hours to be detectable, and in some cases days to be complete.

2B Types of mechanism

In spite of this wide diversity, there is one feature which is common to the mechanism of action of all hormones. In all cases there is a need for a transfer of information across the plasma membrane, from the hormone in the circulation to the inside of the cell where the final response will take place.

There are two basic types of mechanism for achieving this information transfer. In the first, the hormone itself enters the cell and binds to a recognition site on an intracellular receptor protein. In the second type of mechanism, the hormone binds to a recognition site on the surface of the cell and does not itself enter the cell immediately.

2B1 Hormones which enter the cell

Steroid hormones function by the first type of mechanism. Steroids are small lipid-soluble molecules which pass freely across the plasma membrane. In doing so, the need for a transfer of information across the plasma membrane is satisfied. The hormone binds to an intracellular protein receptor and the hormone/receptor complex itself has direct effects on the metabolism of the cell. Thyroid hormones also have intracellular receptors which are thought to be located in the nucleus.

2B2 Hormones which interact with the cell surface: The second messenger hypothesis

If the hormone binds first to the surface of a cell and does not enter the cell, the mechanism for transfer of information across the cell membrane is more complicated. This aspect of hormone action was first defined by Sutherland (1959) in the second messenger hypothesis. The hormone is the first messenger which binds to a specific recognition site on the outer surface of the plasma membrane. As a result of this binding, a regulator molecule, the second messenger, changes in concentration in the cytoplasm. It is the second messenger which directly affects intracellular processes. To identify a second messenger for a hormone, a number of criteria must be satisfied:

1 The concentration of the second messenger in the cytoplasm must rise or fall in response to the binding of the hormone to the cell-surface recognition site.
2 The second messenger must affect the metabolism of the cell in a way which is consistent with the physiological effects of the hormone.
3 There must be a mechanism for removing the second messenger and terminating the signal.
4 Ultimately, the sequence of events, addition of hormone, leading to a change in the concentration of the second messenger, followed by the physiological response, must be established. In other words, it is not enough to simply establish the first three conditions, a causal relationship between them must also be demonstrated.

The second messenger hypothesis has provided a general working model for the mechanism of action of all hormones which have their initial interaction with the cell surface. It is also applicable to many other types of transmembrane signalling process.

All hormones whose action is mediated through a second messenger, first bind to a highly specific recognition site on the outer surface of the cell. The binding site is on an integral membrane protein known as the hormone receptor. As a consequence of binding the hormone, the receptor acquires the ability to alter the concentration of the second messenger inside the cell.

2C Relationship of binding to response

The binding of a hormone to the cell surface, or to purified plasma membranes, is relatively easy to measure, and many attempts have been made in many different systems to relate hormone binding to the physiological response of the cell to the hormone. Two theories were suggested in the 1930s to explain drug action: the occupancy theory and the rate theory. Applied to hormones, the occupancy theory would suggest that the response is directly proportional to the number of receptors occupied. The rate theory suggests that one occupation event produces a discrete quantum of response as a single event and that a prolonged response requires further new occupation events. In general, the occupancy theory seems to apply for most hormones. At first sight this suggests that when half the hormone receptors are occupied, half the maximal response should be observed, and in the 1960s evidence was presented in some systems which was supposed to show precisely that. However, the situation is more complex. The relationship of hormone-receptor binding to the response will depend upon the nature of the response measured. If we are dealing with the molecular event which follows immediately after the binding of the hormone to the receptor, then it is likely that this event will take place in proportion to the occupancy of the receptor. However, if there are intervening molecular events, then there is no reason to expect a direct relationship of the response to the hormone binding. The greater the number of intervening steps the greater the difference may be.

The reasons for this are quite simple. One binding event may lead to the generation of numerous molecules of second messenger. Different

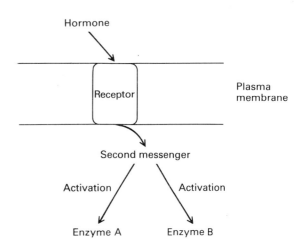

Fig. 2.1. Sequence of events between hormone binding and response. The second messenger activates both enzyme A and enzyme B. If the concentration of second messenger required for half-maximal activation is lower for A than for B, then it follows that a lower hormone-receptor occupancy will also be required.

responses to the hormone may be maximally stimulated by different concentrations of the second messenger. As a result, different responses to the hormone will require different amounts of hormone receptor to be occupied. If the very simple situation described in Fig. 2.1 is considered, the second messenger concentration and the activities of enzymes A and B can be measured. The rate of production of the second messenger depends upon the number of hormone receptors occupied by the hormone. However, activation of enzyme A may require a low concentration of the second messenger, and therefore a low receptor occupancy, while activation of enzyme B may require a high concentration of the second messenger and, accordingly, a much higher receptor occupancy. The situation is usually much more complicated, with several molecular events between the binding of the hormone to its receptor and the generation of the second messenger, and also between the generation of the second messenger and the final change in enzyme activity.

In many cases, the final metabolic response of cells to a hormone requires that a relatively small proportion of the receptors be occupied. This has advantages to the cell. A hormone that binds with a simple binding reaction with no cooperativity will give a straightforward hyperbolic binding curve (Fig. 2.2). In order to increase the number of receptors occupied from 50% to around 100%, the concentration of

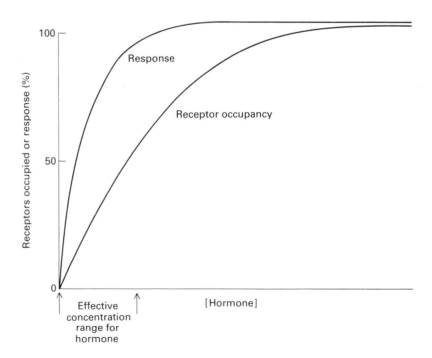

Fig. 2.2. Relationship of receptor occupancy to physiological response.

hormone needs to be increased tenfold. However, at low occupancies the relationship between the concentration of the hormone and the number of receptors bound is close to linear. Under these conditions a twofold increase in occupany allows a twofold increase in response. Thus operating at low degrees of receptor saturation renders the physiological response more sensitive to changes in hormone concentration, and it is advantageous for the maximal response to take place when only a small proportion of the receptors are occupied.

2D Mechanisms for second messenger generation

There are a number of possible mechanisms by which the concentration of the second messenger inside the cell can be altered.

1 The receptor may alter the activity of a membrane enzyme with its active centre on the inner surface of the plasma membrane. Since the concentration of the second messenger can either increase or decrease, the receptor may activate or inhibit either the synthesis or the breakdown of the second messenger. In the case of adenosine 3',5'-cyclic monophosphate (cyclic AMP), many hormones have been shown to increase its rate of synthesis, some hormones have been shown to decrease its rate of synthesis, and it has been proposed that some hormones increase its rate of breakdown.

2 The binding of the hormone to a receptor may alter a specific transport process across the cell membrane. This type of mechanism has usually been proposed when the second messenger is an inorganic ion and there is a substantial difference in concentration between the cytoplasm and the extracellular space. Both Ca^{2+} and Na^+ ions have been suggested to act as second messengers for hormones. The hormone may change the rate of influx or efflux of an ion down the concentration gradient, or alter the rate of active transport of an ion against the concentration gradient.

3 The binding of the hormone to the receptor may lead to the release of a second messenger from binding sites on the inner surface of the plasma membrane. It has been proposed that bound Ca^{2+} may be released in this way. In the case of insulin, it has been suggested that a peptide which may be a component of the hormone receptor is released from the inner surface of the plasma membrane.

4 The binding of the hormone may lead to the internalization of the hormone–receptor complex. This happens in the case of growth factors. The process is too slow to account for the rapid effects of growth factors, but the internalized receptor could play a role in the long-term response.

There is some evidence for all these types of mechanism. Most rapid hormone responses appear to use either cyclic AMP or Ca^{2+} as a second messenger. Many hormones employ both, using different receptor types for each response. Furthermore, the two second messengers interact in their effects on metabolism. In some systems they antagonize each other's response, while in other systems or under different circumstances the response to the other messenger may be reinforced.

Some hormones do not appear to function via changes in cyclic AMP or Ca^{2+} in a straightforward manner, the most important example being insulin. It should also be stressed that the cell does not respond passively to the hormone. The nature of the response will be governed to some extent by the metabolic state of the cell and there are a variety of feedback mechanisms whereby the hormone response may be modulated. In the following chapters these questions will be considered in detail.

Chapter 3 Regulation of cyclic AMP concentration by hormones

3A Introduction: cyclic AMP and the second messenger hypothesis

The discovery of cyclic AMP by Sutherland and his colleagues in the late 1950s led to an understanding of the mode of action of a large class of hormones. Before this the mechanisms by which hormones produced their effects upon target cells were largely obscure. Sutherland found that cyclic AMP increased the rate of glycogen breakdown in liver homogenates, and that the reason for the increase in rate was an increase in the activity of glycogen phosphorylase. They also showed that the plasma membrane contained an enzyme, adenylate cyclase, which was activated by adrenaline and which catalysed the conversion of ATP to cyclic 3',5'-AMP. A second enzyme activity, cyclic AMP phosphodiesterase, was discovered which converted cyclic AMP to 5'-AMP. These observations led to the formulation of the second messenger hypothesis. Specifically related to cyclic AMP, the hypothesis proposed the following sequence of events.

1 The hormone binds to a protein receptor with a specific recognition site on the outer surface of the plasma membrane.

2 The binding of the hormone leads to a conformational change which results in the activation of adenylate cyclase. The active site of adenylate cyclase is located on the inner surface of the plasma membrane, so that

Table 3.1. Receptor-mediated effectors of adenylate cyclase

Hormone	Activation or Inhibition
β-Adrenergic	+
α_2-Adrenergic	−
Glucagon	+
Secretin	+
ACTH	+
TSH	+
Vasopressin	+
5-Hydroxytryptamine	+
Opiates	−
Dopamine	+
Prostaglandins	±
Adenosine	±

activation leads to increased conversion of cytoplasmic ATP to cyclic AMP resulting in a rise in the cytoplasmic cyclic AMP concentration.
3 Cyclic AMP acts as an effector to modulate the metabolism of the cell.
4 As the circulating hormone level drops, the activity of adenylate cyclase drops, and the cytoplasmic cyclic AMP concentration falls as a consequence of the presence of cyclic AMP phosphodiesterase.

Over the years, hormone-sensitive adenylate cyclases have been found in almost all the vertebrate cell types which have been studied, the major exception being the human erythrocyte. A large number of hormones and other effectors use this mechanism. The list in Table 3.1 is by no means exhaustive but should give an idea of the extent of the diversity. The second messenger hypothesis has been shown to be essentially correct with respect to cyclic AMP, although the details of the mechanisms involved have turned out to be much more complicated. This chapter deals with the regulation of cyclic AMP concentration in the cell. The mechanisms by which cyclic AMP alters metabolism by the activation of protein phosphorylation are considered in Chapter 4.

3B The adenylate cyclase reaction

Adenylate cyclase catalyses the conversion of ATP to cyclic 3',5'-AMP with the elimination of pyrophosphate (Fig. 3.1). In most tissues the pyrophosphate is rapidly hydrolysed to inorganic phosphate by the enzyme pyrophosphatase. This has the effect of increasing the overall negative free energy change of the reaction so that it becomes effectively irreversible under the conditions prevailing in the cell (see Chapter 1). The substrate for adenylate cyclase is $MgATP^{2-}$, and the enzyme will also use $MnATP^{2-}$. There is also a requirement for free Mg^{2+} or Mn^{2+}. Since the concentration of Mg^{2+} in the cell is much higher than the concentration of Mn^{2+}, it seems likely that under physiological conditions only Mg^{2+} will be relevant.

The K_m of adenylate cyclase for $MgATP^{2-}$ in most vertebrate tissues has been found to be between 50 and 100 µmol/l. Intracellular concentrations of ATP vary between different tissues but are usually in the range 5–10 mmol/l. Therefore, in a normal cell there is always enough ATP to allow the enzyme to operate at V_{max}. The concentrations of cyclic AMP required to produce an effect are between 1 and 10 µmol/l. The enzyme activity needed to generate this is very low in relation to the other ATP demands of the cell, so that even when adenylate cyclase is maximally activated, the rate of utilization of ATP is small in relation to the total ATP turnover. Thus the activity of adenylate cyclase is never

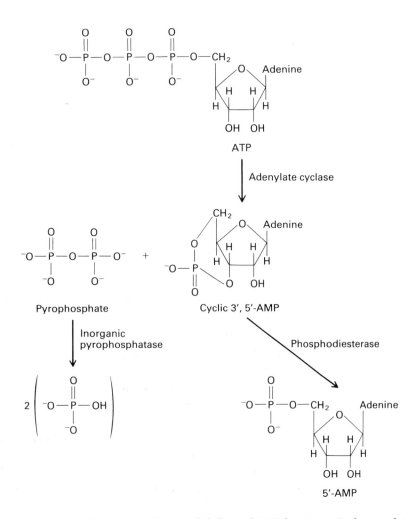

Fig. 3.1. Synthesis and breakdown of cyclic 3′,5′-AMP.

limited by changes in the availability of ATP but is entirely regulated by specific effectors. Conversely, the activation of adenylate cyclase never reduces the cytoplasmic ATP concentration, and the consequences of activation are confined to the specific effects of cyclic AMP on the metabolism of the cell.

3C Phosphodiesterases

Cyclic AMP phosphodiesterases catalyse the reaction, cyclic AMP to 5′-AMP (Fig. 3.1), and are the enzymes responsible for removing cyclic AMP from the cytoplasm. There are also phosphodiesterases which catalyse the hydrolysis of cyclic GMP. It has proved to be surprisingly difficult to purify phosphodiesterases and determine their properties. In some reports it was suggested that mammalian cells may contain as many as six or seven different phosphodiesterases using cyclic 3′,5′-AMP as substrate. Later it became apparent that many of these apparently distinct species were produced as a result of proteolytic damage during extraction of the tissue.

It is now clear that cells contain three basic types of phosphodiesterase. The first has a low K_m for cyclic GMP, a high K_m for cyclic AMP, and a relatively low V_{max}. The second type of enzyme has a high K_m for both cyclic AMP and cyclic GMP, and a relatively high V_{max}. High-K_m enzymes occur as integral membrane enzymes in the plasma membrane and also as soluble enzymes in the cytoplasm. They may serve to impose an upper limit on the concentration to which cyclic 3′,5′-AMP may rise since the V_{max} is high; in other words they may act as a sort of safety valve. The soluble form is activated by Ca^{2+} acting through calmodulin, the effect being to reduce markedly the K_m for cyclic AMP and also for cyclic GMP (see Section 7E1). When activated by calmodulin, this enzyme may be active at the levels of cytoplasmic cyclic AMP produced in response to hormones.

The third enzyme has a low K_m for cyclic AMP of about 1μmol/l, a relatively low V_{max}, and is probably the enzyme responsible for the conversion of cyclic AMP to 5′-AMP under normal conditions. This enzyme appears to be partially associated with the plasma membrane and with other cell membranes as a peripheral membrane enzyme, and partially soluble in the cytoplasm. The plasma membrane-associated form of the enzyme has recently been shown to be activated by insulin in liver by a complex mechanism which will be described in Chapter 10.

A number of drugs act by their ability to inhibit cyclic AMP phosphodiesterase. This tends to increase cyclic AMP levels in the cell and also increases the rise in cyclic AMP in response to a hormone. The most familiar of these drugs is caffeine, probably the most widely used stimulant in the world, which acts as a result of its ability to increase cellular cyclic AMP levels.

3D Regulation of adenylate cyclase: the role of guanine nucleotides

It is now well established that the activation of adenylate cyclase by hormones requires guanine nucleotides. This was first suggested by Rodbell and his colleagues (1971) in the course of studies of the binding of glucagon to purified liver plasma membranes. They noticed that GTP promoted the dissociation of the hormone. The extent of activation of adenylate cyclase by the hormone appeared to be dependent upon the concentration of the substrate, MgATP^{2-}, in the enzyme assay. As the substrate concentration increased from µmol/l to mmol/l levels, the extent of activation by glucagon was increased (Fig. 3.2). They also found that the addition of low concentrations of GTP of the order of 1 µmol/l affected adenylate cyclase activity. GTP activated adenylate cyclase and also increased the extent of the activation by glucagon regardless of the ATP concentration (Fig. 3.2). The effect was synergistic in that the activity in the presence of both glucagon and GTP was

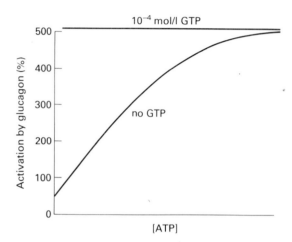

Fig. 3.2. Effect of ATP concentration on the stimulation of liver adenylate cyclase by glucagon in the presence and absence of GTP. Upon the addition of GTP, activation by glucagon becomes independent of the ATP concentration.

greater than the sum of the activities in the presence of glucagon, or of GTP, alone (Table 3.2). Later it was established that the ATP available from commercial suppliers at that time was slightly contaminated with GTP and that, since very low levels of GTP are required for the response, this contamination was enough to produce an effect. If the ATP used as substrate in the adenylate cyclase assay was carefully purified to remove contaminating GTP, the activation of adenylate cyclase by the hormone became almost completely dependent upon the addition of GTP.

Table 3.2. Relative activities of rat liver plasma membrane adenylate cyclase in the presence of glucagon and guanine nucleotides

Basal	1
GTP	3
Glucagon	2
Glucagon + GTP	10
p(NH)ppG	10

The next question to be considered is the mechanism of the involvement of GTP. An obvious possibility was that the hydrolysis or transfer of the terminal phosphate might be needed for activation of adenylate cyclase. This was examined using an analogue of GTP, guanylyl imidodiphosphate, p(NH)ppG (Fig. 3.3). The dimensions of this compound are almost identical to those of GTP, but the substitution of a $-NH-$ group for $-O-$ in the linkage of the terminal phosphate makes it resistant to hydrolysis or transfer. If the activation of adenylate cyclase by GTP

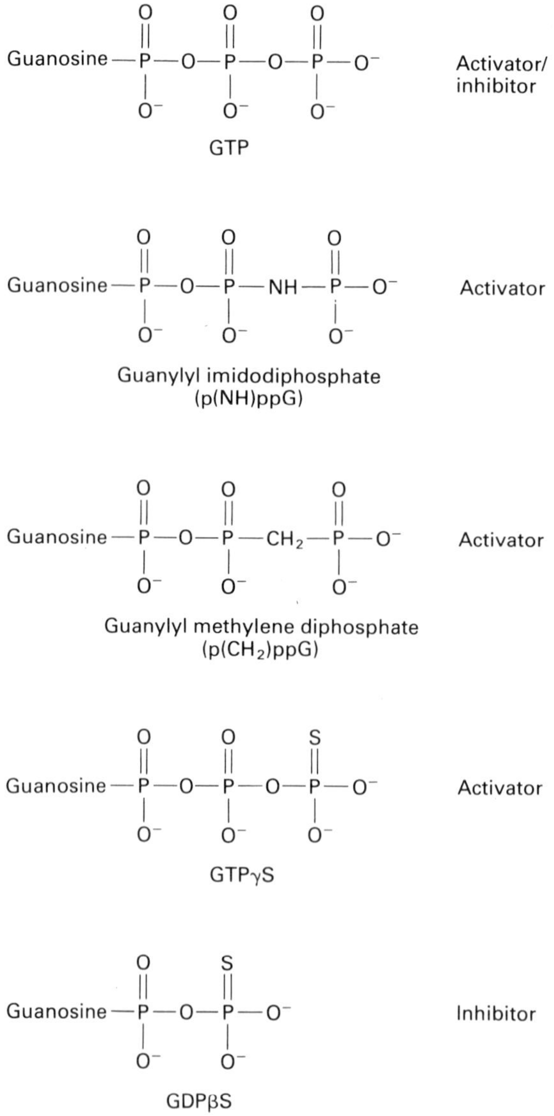

Guanosine—P—O—P—O—P—O⁻ ... Activator/inhibitor

GTP

Guanosine—P—O—P—NH—P—O⁻ ... Activator

Guanylyl imidodiphosphate
(p(NH)ppG)

Guanosine—P—O—P—CH$_2$—P—O⁻ ... Activator

Guanylyl methylene diphosphate
(p(CH$_2$)ppG)

Guanosine—P—O—P—O—P—O⁻ ... Activator

GTPγS

Guanosine—P—O—P—O⁻ ... Inhibitor

GDPβS

Fig. 3.3. Structures of GTP and GDP analogues used in adenylate cyclase studies.

requires the cleavage of the terminal phosphate bond, then the analogue should be less effective as an activator. However, it was found that in liver plasma membranes, p(NH)ppG was much more effective as an activator of adenylate cyclase than GTP (Table 3.2), and that the activation was essentially irreversible. The addition of glucagon did not increase the eventual activation in the presence of p(NH)ppG but the maximum activity was achieved more rapidly. Other analogues of GTP, p(CH)ppG and GTPγS (Fig. 3.3), have been found to have the same effect. Based on these observations, Rodbell suggested that the adenylate cyclase system contained three functional sites. The binding of a hormone to a recognition site led to an increase in the association of GTP with a second site that he called the transducer. The binding of GTP to the transducer site promoted the activation of the third site, the catalytic site of the enzyme. Thus the activator of the enzyme is GTP and the effect of the hormone is to increase the effectiveness of GTP as an activator rather than to affect directly the catalytic activity itself. The apparently irreversible nature of the activation by p(NH)ppG led to the suggestion that the activation of the enzyme by GTP was reversed as a consequence of the hydrolysis of GTP to GDP and P_i (Fig. 3.4).

If the cycle of activation and inactivation described in Fig. 3.4 is correct, adenylate cyclase should contain a site which has GTPase activity in addition to the hormone receptor site and the catalytic site. The GTPase activity should also be affected by hormones which activate adenylate cyclase. The hypothesis was that the activated state of the

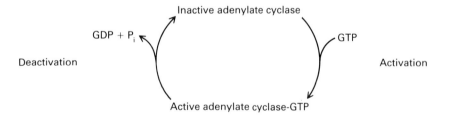

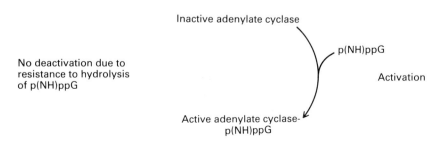

Fig. 3.4. Activation/deactivation cycle with GTP and p(NH)ppG.

enzyme contains bound GTP and that the effect of the hormone is to increase the proportion of the enzyme in this state. There are two ways in which this could be achieved. The hormone might inhibit the hydrolysis of GTP to GDP and P_i thus increasing the lifetime of GTP on the GTP binding site. Alternatively, the hormone might increase the rate of binding of GTP to the enzyme. This would also increase the effectiveness of GTP as an activator, but in this case, since the concentration of the enzyme–GTP complex increases, the rate of hydrolysis of GTP should also increase. The analogue, p(NH)ppG, cannot be hydrolysed, but the rate at which it activates adenylate cyclase was increased by glucagon. This suggested that it was the binding step which was activated by the hormone, but did not preclude the possibility that the GTPase activity might be inhibited as well.

Attempts to measure the GTPase in rat liver plasma membranes were unsuccessful because the non-specific GTPase activity, unrelated to the regulation of adenylate cyclase, was very high and it was not possible to measure the specific activity against this background. Turkey erythrocyte plasma membranes contain adenylate cyclase which is activated by β-adrenergic hormones. In this case it was possible to detect a GTPase which was activated by adrenaline, suggesting that the effect of hormones is to promote the association of GTP with the GTPase. The effect of inhibition of the GTPase could also be examined using cholera toxin which was known to cause a GTP-dependent irreversible activation of adenylate cyclase. The effect of the toxin was to make the response to GTP similar to the activation by p(NH)ppG, in that the extent of activation was increased and the reversal of activation was slower. The

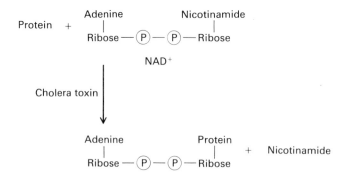

Fig. 3.5. The cholera toxin ADP-ribosylation reaction.

cholera toxin effect required NAD. It had been suggested that, in common with other bacterial toxins, cholera toxin produced its effects as a result of a covalent modification of adenylate cyclase using NAD to ADP-ribosylate the enzyme (Fig. 3.5). It was found that cholera toxin inhibited the GTPase provided that NAD was present. The toxin also catalysed the incorporation of ^{32}P-ADP ribose from ^{32}P-NAD into a protein which had a molecular weight of 42 000. The labelling was dependent upon the presence of GTP, and the same protein could be radiolabelled with a photoreactive analogue of GTP suggesting that it was a GTPase.

The model outlined in Fig. 3.4 is well supported. Adenylate cyclase contains a GTPase site. The association of GTP, or analogues of GTP, with the site activates adenylate cyclase and is enhanced by hormones. The hydrolysis of GTP to GDP and P_i reverses the activation, and the inhibition of this GTPase activity by cholera toxin leads to persistent activation by GTP.

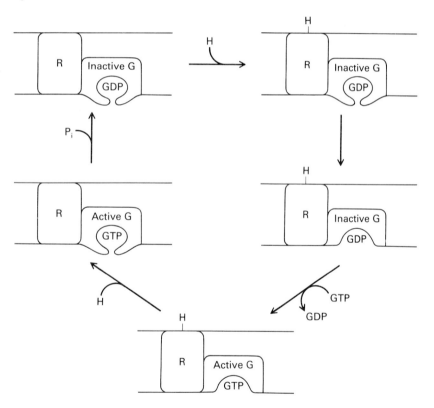

Fig. 3.6. Model for the effect of hormone on guanine nucleotide exchange. R=Hormone receptor. H=Hormone. G=Guanine nucleotide binding protein.

There are a number of other advantages in using turkey erythrocyte plasma membranes to study the interaction between hormones and guanine nucleotides in the activation of adenylate cyclase. In mammalian systems there is substantial activation with GTP alone and the hormone simply enhances this activation. In the turkey erythrocyte, activation of adenylate cyclase by GTP is completely dependent upon the presence of a β-adrenergic hormone. The activation by p(NH)ppG also requires the presence of hormone in the turkey erythrocyte, while in mammalian plasma membranes, hormone increases the rate but not the extent of activation by p(NH)ppG. Obviously, since the effect of hormone on the guanine nucleotide activation is more pronounced, it is easier to study. A second advantage is the hormone involved. Turkey erythrocyte adenylate cyclase is activated by adrenaline acting through a β receptor. There are a number of drugs available known as β blockers which compete with the hormone for binding to the receptor. This makes it possible to reverse the hormone effect more or less instantaneously. This is very difficult in the case of a polypeptide hormone such as glucagon where the only method of reversal requires extensive washing of the membranes.

Further experiments using turkey erythrocyte plasma membranes revealed that the GTP binding site contains bound GDP in the inactive state. The effect of a hormone acting via its receptor is to cause a conformational change which allows the bound GDP to exchange for

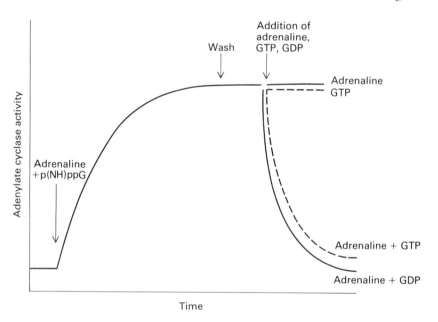

Fig. 3.7. Reversal of adenylate cyclase activation in turkey erythrocyte plasma membranes by GTP, GDP and adrenaline. Adenylate cyclase was activated by the addition of adrenaline and p(NH)ppG. After completion of activation, free adrenaline and p(NH)ppG were removed by washing. Reversal of the activation required the addition of either GTP or GDP together with the hormone. Adrenaline or guanine nucleotide alone had no effect.

another guanine nucleotide. Under physiological conditions this will be GTP. As long as the site contains GTP, adenylate cyclase will be activated. When GTP is hydrolysed by the GTPase activity of the guanine nucleotide binding protein leaving GDP on the binding site, activation is reversed. The enzyme can only be reactivated by the addition of hormone together with GTP or an activatory analogue of GTP such as p(NH)ppG which exchange with the bound GDP (Fig. 3.6).

Two experiments support this model. Turkey erythrocyte plasma membrane adenylate cyclase was activated with adrenaline and p(NH)ppG, and the hormone and unbound p(NH)ppG removed by washing the membranes. Adenylate cyclase stayed in the active state since p(NH)ppG cannot be hydrolysed. To reverse the activation it was necessary to add both hormone, to open the site, and GTP or GDP, to exchange for the bound p(NH)ppG (Fig. 3.7). In a second experiment the membranes were incubated with adrenaline together with GTP which was radioactively labelled in the guanine ring (Fig. 3.3). The activation process was stopped by first adding a β blocker to prevent hormone action and then by washing the membranes to remove unbound labelled GTP. The model predicts that the addition of unlabelled GTP or p(NH)ppG, together with fresh hormone, should result in the release of radioactively labelled GDP. This was the result observed.

3E Protein components of adenylate cyclase

From the discussion above it is clear that the activation of cyclic AMP synthesis by hormones requires at least three distinct functional sites: a recognition site for the hormone, a guanine nucleotide binding site and a catalytic site. The next question to consider is how these sites are physically organized in relation to each other. The simplest possible arrangement is shown in Fig. 3.8. This consists of a single protein spanning the plasma membrane with the recognition site for hormones on the outer surface and the catalytic site and guanine nucleotide regulatory site on the inner surface. The hormone would then act in the same way as any other effector binding to an enzyme, causing a conformational change and leading to an increase in enzyme activity. The only complicating factors are the need for a second effector, the guanine nucleotide and the asymmetric arrangement of the enzyme in the plasma membrane which allows for a transfer of information from the outside of the cell to the inside of the cell. In fact it appears that each functional site is located on a separate protein unit, usually referred to as R for the hormone receptor, G, or by some authors N, for the guanine nucleotide binding component, and C for the catalytic component.

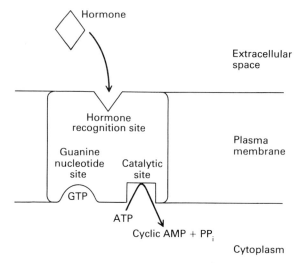

Fig. 3.8. Single transmembrane protein model for adenylate cyclase. In this early model adenylate cyclase functions as an allosteric enzyme in which the catalytic site is on the inner surface of the plasma membrane, one regulatory site for the hormone is on the outer surface, and the other regulatory site for guanine nucleotides is on the inner surface.

3E1 The guanine nucleotide binding protein

It is clear that the guanine nucleotide binding site and the catalytic site of adenylate cyclase are on separate distinct proteins. This first became apparent from studies of a cultured mouse cell line, the S49 lymphoma cell. The parent cell line contains an adrenaline-activated adenylate cyclase. A derivative of this cell line, cyc⁻, has been isolated in which purified plasma membranes apparently contained no adenylate cyclase activity. Binding studies showed that the hormone receptor was present. It was known from other systems that Mn^{2+} is able to activate adenylate cyclase by a direct effect on the catalytic unit independently of hormone or guanine nucleotides. In the presence of Mn^{2+}, the adenylate cyclase activities of cyc⁻ cell plasma membranes and of membranes from the parent S49 lymphoma cell line were similar. It appeared that the enzyme had lost the ability to respond to guanine nucleotides. Since hormones act by potentiating the guanine nucleotide response, this will lead to the loss of the hormone response as well. This conclusion was reinforced by the fact that cholera toxin had no effect on the activity of cyc⁻ adenylate cyclase. The protein of 42 000 molecular weight which was ADP-ribosylated by cholera toxin and covalently labelled by photoreactive GTP analogues, was also absent. Later it was found that the guanine nucleotide response of cyc⁻ S49 membranes could be restored by extracts from membranes in which the adenylate cyclase did respond to guanine nucleotides. This has provided a means of detecting the presence of the G protein in cell extracts and, as a result, considerable progress has been made towards its purification.

3E2 The hormone receptor

The model outlined in Fig. 3.8, in which the hormone receptor and the catalytic unit of adenylate cyclase form a single transmembrane complex, was rapidly shown to be inadequate. In many cell types, more than one hormone was able to activate adenylate cyclase, the most striking example being the rat fat cell where as many as seven different hormones activate the enzyme. This raises two possibilities: either a single receptor is able to recognize several different hormones, or there must be several different hormone receptors. The structure of the hormones involved varied greatly from the catecholamine, adrenaline, at one extreme, to ACTH, a large polypeptide, at the other extreme. It seemed unlikely, therefore, that there was a common receptor for all the hormones. This was tested by exposing intact fat cells to the proteolytic enzyme, trypsin, so that only cell-surface proteins would be affected. This led to a complete loss of the ability of glucagon to activate adenylate cyclase while the response to adrenaline was unaffected. The response to other hormones was affected to varying extents. It seemed likely, therefore, that each hormone had a separate receptor protein. This raised the possibility that each different type of hormone receptor had its own distinct population of adenylate cyclase catalytic units, that is, one population activated by Hormone A and a quite separate population by Hormone B. In that case the activation by different hormones should be additive. In fact if adenylate cyclase was activated to the maximum extent by the most effective hormone, then the presence of other hormones had no additional activating effect. This led to the conclusion that several different types of hormone receptor protein compete for a common pool of adenylate cyclase catalytic units.

3F The mobile receptor hypothesis

There are two possible physical arrangements by which several different types of receptor can compete for a common pool of catalytic units. The catalytic unit might have a ring of receptors clustered around it and permanently in contact. This can be called the static model (Fig. 3.9a). The second possibility is described by the mobile receptor hypothesis (Fig. 3.9b). This suggests that the hormone receptor and catalytic unit are separate from each other but are able to move relative to each other in the fluid matrix of the plasma membrane. When the receptor is occupied by hormone, its affinity for the catalytic unit is increased so that when the receptor makes contact with the catalytic unit, it binds to and activates it. The activation requires GTP as described above.

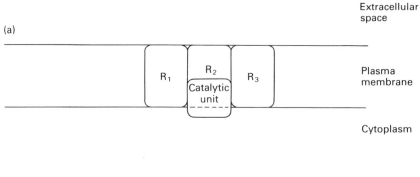
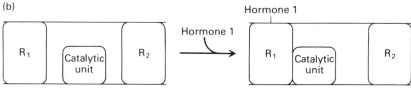

Fig. 3.9. The static and mobile receptor models. (a) Static precoupled model. Receptors and catalytic unit precoupled as a single unit in the absence of hormone. (b) Mobile receptor model. Receptors and catalytic unit separate in the absence of hormone. Binding of hormone to the receptor leads to its association with the catalytic unit.

Convincing support for the mobile receptor hypothesis came from cell fusion studies by Schramm and his colleagues (1976). Sendai virus has the ability to promote the fusion of cultured tumour cells with normal cells. The two cells form a common plasma membrane and fluorescent labelling experiments show that the plasma membrane proteins from both cells mix evenly in the fluid-lipid matrix (Fig. 3.10). Two cell lines were chosen, a Friend erythroleukemia cell which contained adenylate cyclase but no β-adrenergic hormone receptor, and a turkey erythrocyte which contained both adenylate cyclase and a β receptor. The turkey erythrocytes were treated with the sulphydral reagent N-ethyl maleimide (NEM). This irreversibly inhibited adenylate cyclase without affecting the ability of the β receptor to bind hormone. Thus they now had two types of cell: one containing a functional β receptor but no catalytic unit, and one containing adenylate cyclase catalytic activity but no β receptor. The cells were fused and the plasma membrane fraction isolated. The adenylate cyclase in membranes from fused cells was activated by adrenaline. This demonstrated unequivocally that the hormone receptor was able to move in the lipid matrix of the plasma membrane to activate adenylate cyclase, and was convincing

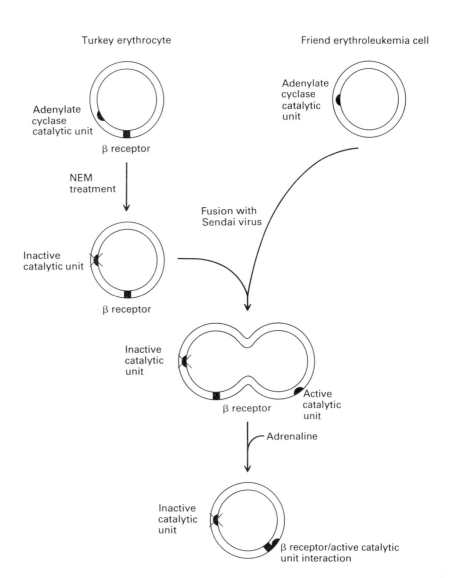

Fig. 3.10. Cell fusion experiment to demonstrate receptor mobility.

evidence in support of the mobile receptor hypothesis. Later, the same result was achieved using cells from a variety of different sources and a variety of different fusion techniques. It is possible to activate adenylate cyclase from one species with a receptor obtained from a different species. This has the interesting implication that the protein structures involved in the interaction between the hormone receptor and the catalytic unit must have been very highly conserved during evolution.

3G The collision coupling model

The static model and the mobile receptor model can be described in terms of the equations in Fig. 3.11. These equations describe the situation in the presence of p(NH)ppG. In this case the effect of the hormone will be to promote the activation of the enzyme, and deactivation will not occur since the guanine nucleotide cannot be hydrolysed. Thus in the presence of saturating concentrations of p(NH)ppG, the kinetics of the activation in response to the hormone can be studied in isolation from the deactivation process. This is particularly easy in turkey erythrocyte membranes where activation is almost completely dependent upon the presence of hormone and is also relatively slow making it simple to follow experimentally. Finally, since the activation is via a β receptor, the activation process can be stopped instantaneously with a β blocker.

Tolkovsky and Levitzki (1978) undertook a theoretical consideration of the kinetics which would be predicted by the two different models in this specific situation where reversal of activation was blocked. They concluded that the static model would give first-order kinetics with respect to the increase in activity in response to the hormone, and no

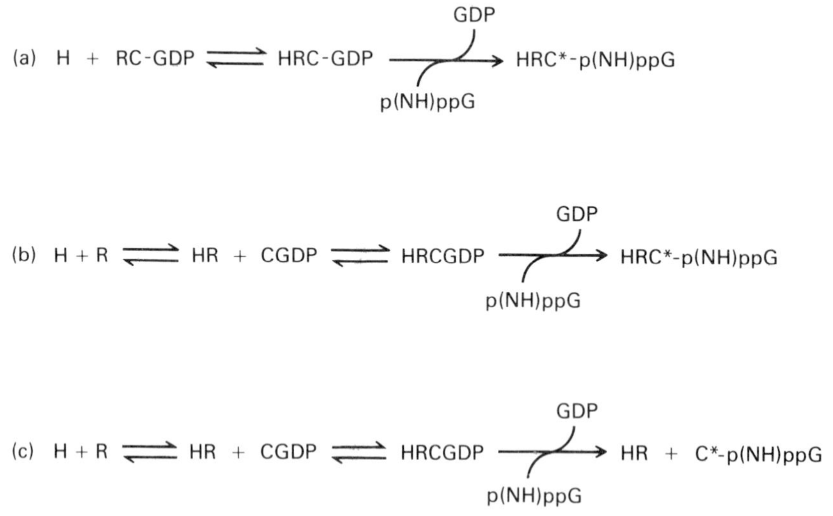

(a) H + RC-GDP ⇌ HRC-GDP —[GDP / p(NH)ppG]→ HRC*-p(NH)ppG

(b) H + R ⇌ HR + CGDP ⇌ HRCGDP —[GDP / p(NH)ppG]→ HRC*-p(NH)ppG

(c) H + R ⇌ HR + CGDP ⇌ HRCGDP —[GDP / p(NH)ppG]→ HR + C*-p(NH)ppG

Fig. 3.11. Different models for the activation of adenylate cyclase by hormones and p(NH)ppG. (a) Static model (precoupled receptor and catalytic unit). (b) Mobile receptor model. (c) Collision coupling model. H=hormone, R=hormone receptor, C=catalytic unit, C*=activated catalytic unit. In all cases the final activation involving p(NH)ppG is irreversible. For simplicity, the existence of a separate guanine nucleotide binding protein is ignored.

cooperativity in the hormone binding. The mobile receptor model would give rise to deviations from first-order kinetics and negative cooperativity in the hormone binding. They then examined the kinetics of activation of turkey erythrocyte plasma membrane adenylate cyclase by adrenaline in the presence of saturating concentrations of p(NH)ppG. They found no cooperativity in the hormone binding and first-order kinetics. This supported the static model conflicting with the evidence from cell fusion studies which favoured the mobile receptor model. A third model, however, would be expected to give first-order kinetics and non-cooperative hormone binding, while being consistent with the cell fusion experiments (Fig. 3.11c). In this case the hormone−receptor complex forms an association with adenylate cyclase as in the mobile receptor hypothesis. However, as the activation process is completed by the binding of p(NH)ppG, the receptor is released thus becoming available to initiate activation of further units of adenylate cyclase. Thus the association of the receptor is transitory and this mechanism has been described as collision coupling.

Thus there were two models each of which should give rise to first-order kinetics and non-cooperative hormone binding: the static model and the collision coupling model. They could be distinguished by varying the number of hormone receptors. This was done with a β blocker which was able to react covalently and irreversibly with the β receptor. By varying the concentration of the β blocker and the length of time during which the membranes were exposed to it, varying proportions of the β receptors on the membranes would be irreversibly inactivated. The static model predicts that the activated state requires a constant association of an occupied hormone receptor with the catalytic unit. Thus a reduction in the number of receptors should reduce the extent of activation but not affect the rate of activation (Fig. 3.12a). The collision coupling model predicts that after the activation of one adenylate cyclase unit is complete, the hormone receptor is released and becomes available to activate further adenylate cyclase units. Since, in the presence of p(NH)ppG, the activation is irreversible, a reduction in the number of receptors will not affect the maximum activation achieved, but will increase the time needed to achieve maximum activation (Fig. 3.12b). The experimental results were as predicted by the collision coupling model.

The collision coupling model is widely accepted. It appears that the hormone receptor interacts with the guanine nucleotide binding component rather than directly with the catalytic unit. This is to be expected since the hormone receptor appears to function by modulating the interaction of guanine nucleotides with the system. It has also been

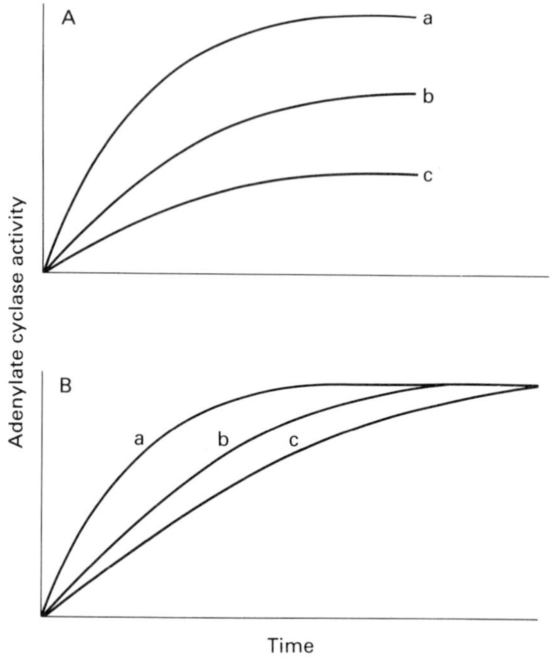

Fig. 3.12. Effect of reduction in receptor number on the activation of adenylate cyclase by adrenaline and p(NH)ppG, predicted by (A) the static precoupled model, and (B) collision coupling. In both cases the rate of approach to complete irreversible activation by p(NH)ppG is followed. From a to c shows the predicted effect of reducing the number of receptors in each model.

found that the binding affinity of hormones for their receptors is affected by GTP, supporting the idea of an interaction between the receptor and the G protein. This raises the question of the nature of interaction of the G protein with the catalytic unit.

It is well established that the G protein and the catalytic unit are separate, distinct protein components, so, as with the hormone receptor, a number of mechanisms can be envisaged for the coupling of the G protein to the catalytic unit. Two models have been most frequently proposed. In one case it is assumed that the G protein and the catalytic unit remain coupled throughout the cycle of activation and deactivation (Fig. 3.13a). The second model suggests that both hormone receptor and the G protein are separate from the catalytic unit in the deactivated state (Fig. 3.13b). The following sequence of events is proposed. The hormone receptor binds to the receptor increasing its affinity for the G protein. The receptor binds to the G protein and promotes the opening of its binding site so that bound GDP can exchange for GTP. On binding GTP, and receptor dissociates leaving an activated guanine nucleotide binding protein which is then able to bind to, and activate, the adenylate cyclase catalytic unit. After the bound GTP is hydrolysed to GDP and P_i, the G protein dissociates to complete the cycle (Fig. 3.13b). There is no conclusive evidence to favour one model over the other. However,

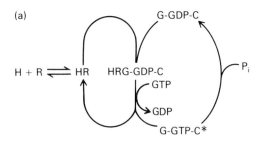

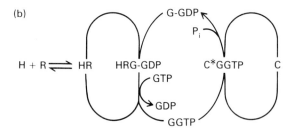

Fig. 3.13. Alternative models for the interaction between the guanine nucleotide binding protein (G) and the catalytic unit (C). (a) Precoupled GC complex. (b) Independent G and C in the absence of GTP. H=hormone, R=hormone receptor, G=guanine nucleotide binding protein, C=catalytic unit, C*=activated catalytic unit. Collision coupling of the hormone/receptor complex (HR) is assumed in both cases.

studies of the kinetics of activation of turkey erythrocyte adenylate cyclase by p(NH)ppG gave no indication of a separation of the G protein from the catalytic unit.

Recently the components involved in the effects of guanine nucleotides on adenylate cyclase have been identified. Three proteins appear to be involved. The first, Gs, has a molecular weight of 42 000 and is the guanine nucleotide binding protein which is ADP-ribosylated by cholera toxin. This mediates activatory effects of GTP. The second, Gi, is a guanine nucleotide binding protein which is ADP-ribosylated by pertussis toxin and mediates inhibitory effects of GTP on adenylate cyclase. It has a molecular weight of 41 000 (see Section 3J). The third component has a molecular weight of 35 000. This component appears to interact with both Gs and Gi. It does not bind GTP and appears to be released from the Gs/catalytic unit complex when activation takes place. It appears to function as a repressor of adenylate cyclase activity.

3H Inhibition of adenylate cyclase by hormones

The fact that some hormones inhibit adenylate cyclase was recognized relatively recently, and compared with the activation process the mechanisms involved are poorly understood. Two types of inhibition have been recognized.

3H1 Desensitization

Prolonged exposure of a cell to high concentrations of a hormone which normally activates adenylate cyclase often leads to a reduction in the response to the hormone. This process has also been called down regulation. The phenomenon is not confined to hormones which act through adenylate cyclase; prolonged exposure of intact cells to many hormones leads to a progressive reduction in the hormone response. It has been most widely studied in relation to adenylate cyclase probably because the basis of the mechanism of activation is well understood. There are four possible types of mechanism.

1 The cell produces an inhibitor of adenylate cyclase as part of the response to the hormone.
2 The cell reduces the number of hormone receptors.
3 The cell impairs the ability to the receptor to activate adenylate cyclase.
4 The cell blocks the adenylate cyclase activity directly.

There are examples of the first three mechanisms, but not of mechanism 4.

3H2 Inhibitor production

When adenylate cyclase is activated in an intact cell, cyclic AMP levels usually rise rapidly to a peak and then decline to a steady state level about three or four times the basal level (Fig. 3.14). This is particularly easy to demonstrate in the isolated fat cell. The sharp initial rise in cyclic AMP in fresh cells on addition of hormone can be prevented by

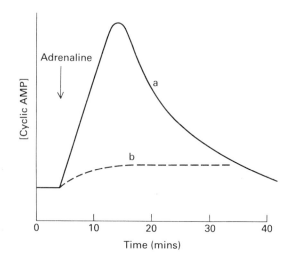

Fig. 3.14. Effect of adrenaline on cyclic AMP levels in fat cells. (a) Adrenaline added to freshly prepared fat cells in fresh medium. (b) Adrenaline added to the same cell preparation as in (a) but suspended in medium from which cells which had been exposed to adrenaline have been removed.

suspending the cells in medium which previously contained cells in which the cyclic AMP rise had been stimulated and allowed to decline. Thus it seems that the cell produces an inhibitor of adenylate cyclase in response to the hormone. This type of effect can only be seen in intact cells and is dependent upon the rise in cyclic AMP. Any factor which raises cyclic AMP levels will block the response to any hormone which works on the cell.

3H3 Removal of receptors

This is a very general process. Prolonged exposure of intact cells to most hormones leads to a reduction in the number of hormone binding sites on the cell surface. This may reflect an actual loss of receptors due to endocytosis of the receptor protein and degradation in the lysosomes. In this case protein synthesis is required for recovery of the hormone binding. In other cases the receptor function is lost but recovers after removal of the hormone by a process which is independent of protein synthesis. In this case the receptor must become ineffective while remaining physically present in the membrane. This type of mechanism is always specific to a particular hormone. In the case of a cell with receptors for more than one hormone, exposure to one hormone does not affect the response to the others.

3H4 Loss of coupling

It has been shown that prolonged exposure of liver plasma membranes to glucagon leads to a loss of the glucagon activation of adenylate cyclase. The effect requires GTP, ATP and divalent cations. These are conditions under which adenylate cyclase is active. However, the concentration of ATP needed is much higher than the K_m for adenylate cyclase, and p(NH)ppA, which is a good substrate for adenylate cyclase, will not substitute for ATP in desensitization. Glucagon binding is not lost, nor is the ability of p(NH)ppG to activate the enzyme. It seems that only the ability of the glucagon receptor to couple to the G protein is lost, and the requirement for ATP suggests that a protein phosphorylation may be involved.

3J Direct inhibition

Many hormones and other effectors, such as prostaglandins, inhibit adenylate cyclase directly. In this case the inhibition is rapid in onset, reversible, and dependent upon a specific receptor. Such inhibitory responses were first reported as early as 1962 but the phenomenon has only been extensively studied since about 1975. Accordingly, mechanisms of inhibition are less well characterized than the activation mechanism.

Inhibitory effects tend to be relatively small; typically there is about a fifty per cent reduction in activity in response to the hormone. In many systems inhibition is potentiated by high concentrations of Na^+ and low levels of free Mg^+. This perhaps explains why they were not recognized for so long, since normal conditions for studying activation of adenylate cyclase are an Na^+-free medium and a large excess of Mg^{2+} over ATP to maximize the proportion of the ATP in the $MgATP^{2-}$ form. However, inhibitory effects have now been demonstrated in a number of different mammalian cell types. The most extensively studied systems are the α_2-adrenergic receptor-mediated inhibitions in hamster fat cell and platelet plasma membranes.

In common with the activation process, hormone-mediated inhibition requires GTP. At first sight this might appear to be a statement of the obvious since the adenylate cyclase activity must first be expressed by activation before it can be inhibited. However, the concentration of GTP required for inhibition is much higher than that needed for activation. Maximum hormone activation is observed at about 10^{-6} mol/l GTP while maximum inhibition requires 10^{-4} mol/l. In plasma membrane preparations which have an inhibitory response, the effect of GTP is usually biphasic (Fig. 3.15). There will be an activatory phase which peaks at about 10^{-6} mol/l GTP, and at higher concentrations an inhibitory phase. The inhibitory phase is intensified by inhibitory hormones. In common with the activatory process, GTP reduces the binding affinity

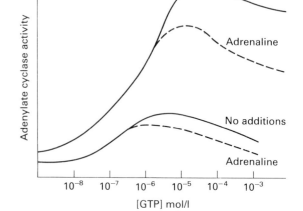

Fig. 3.15. Biphasic response of platelet adenylate cyclase to GTP. Platelet adenylate cyclase responds to prostaglandin E_1 via an activatory receptor and adrenaline via α_2-adrenergic receptors which are inhibitory and act through Gi.

of inhibitory hormones for their receptor, and inhibitory hormones increase GTPase activity. Indeed the GTPase activity is much greater than that seen with activatory hormones. In most cases the hormone-induced inhibition is blocked by analogues such as GTPγS and p(NH)ppG where the terminal phosphate is resistant to hydrolysis, although in some cases it has been reported that the analogues themselves inhibit at high concentrations. It seems likely, therefore, that inhibitory hormone receptors act via the inhibitory guanine nucleotide binding protein, Gi, by a similar mechanism to activation. However, in the case of inhibition, GTP hydrolysis or transfer of the terminal phosphate may be required. The details of the mechanism of action of Gi and of its role in the cell are much less well characterized than the mechanism of action of Gs. Gi also interacts with the 35 000 molecular weight β-subunit. It appears to be present in all mammalian cell membranes, even those where there are apparently no inhibitory hormone receptors. It may have other functions as well as mediating the inhibition of adenylate cyclase.

3K Other effectors of adenylate cyclase

There are many other effectors of adenylate cyclase apart from hormones, guanine nucleotides and divalent cations. Some of these are physiologically important, while others are non-physiological and only of interest if they offer a means of probing the mechanism.

3K1 Physiological effectors

Most naturally occurring effectors other than hormones either activate or inhibit adenylate cyclase by receptor-mediated processes which appear to use essentially the same type of mechanism as hormones. Such effectors include prostaglandins, adenosine, neurotransmitters and encephalins. All these effectors are produced at a site close to the target cell and have a local action on a single tissue in contrast to the general action of a hormone. Apart from this they are very similar to hormones. In the brain, adenylate cyclase is activated by Ca^{2+} via the regulatory protein calmodulin. Calmodulin appears to act directly upon the catalytic unit and not through a G protein. The actions of calmodulin will be considered in detail in Chapter 7.

3K2 Non-physiological effectors

F^- ions. Flouride ions have been known for a very long time to activate adenylate cyclase. The activation is irreversible. Maximum effects

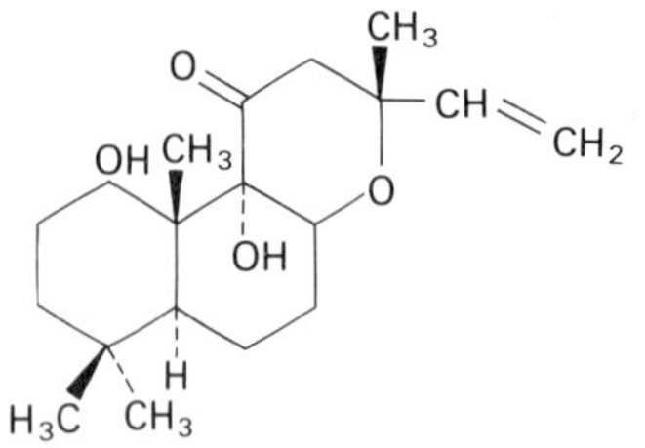

Fig. 3.16. Structure of forskolin.

are seen at about 10 mmol/l and higher concentrations tend to inhibit. The effect requires the activatory G protein since it is not observed in cyc$^-$ S49 lymphoma cells.

Adenosine. In addition to physiological activation or inhibition of adenylate cyclase by adenosine, which occurs in certain specific cell types, there is also a non-physiological inhibition which appears to occur in all cell types. This requires high concentrations of adenosine far in excess of the concentration found in the cell. Analogues of adenosine, in which the purine ring is modified, do not inhibit by this mechanism but are able to act upon the physiological adenosine receptor. On the other hand, analogues with a ribose ring modification are potent inhibitors but the physiological effects are lost.

Forskolin. Forskolin is a diterpene (Fig. 3.16) which activates adenylate cyclase. It is able to activate the catalytic unit very strongly, independently of the G protein and hormone receptors. There may also be an additional component of activation which is dependent upon the activity of G protein.

Suggestions for further reading

Kleinzeller A. & Martin B.R. (Eds) (1983) *Current Topics in Membranes and Transport*, vol. 18, part 1. Academic Press.

Strada S.J. & Thompson W.J. (Eds) (1984) Cyclic AMP phosphodiesterases. *Advances In Cyclic Nucletide Protein Phosphorylation Research*, **16**, section 3. Raven Press.

Advances In Cyclic Nucleotide Research. Raven Press. (Various articles.)

Chapter 4   Mechanism of action of cyclic AMP

4A Introduction

Cyclic 3′,5′-AMP occurs in bacteria and in all animal cells. There is no convincing evidence for its occurrence in plants. In lower life forms it has a number of different mechanisms of action and is involved in the regulation of a wide variety of different processes. In mammals its major role is as a second messenger for hormone action and there appears to be only one mechanism of action, the activation of protein phosphorylation by the cyclic AMP-dependent protein kinase.

The cyclic AMP-dependent protein kinase was the first protein kinase to be discovered that was capable of phosphorylating a number of different proteins, and as a result it is sometimes called the general protein kinase. Recently it has become apparent that the cell contains many different protein kinases controlled in a variety of ways, and that protein phosphorylation is a very widespread mechanism for the operation of control processes.

4B Specificity of cyclic AMP-dependent protein kinase

At first the cyclic AMP-dependent protein kinase was thought to have a relatively little substrate specificity since it will phosphorylate most proteins if a sufficiently high concentration of the kinase and ATP is used. Later the phosphorylation reaction was more completely described. It became apparent that the enzyme does phosphorylate many different proteins and in doing so regulates their activity. A list is given in Table 4.1 which is by no means complete. However, while cyclic AMP-dependent protein phosphorylation is a rather widespread mechanism of regulation, the phosphorylation reaction itself shows quite specific substrate requirements. For example, in the case of phosphorylase kinase, only two serine residues out of a total of 200 are phosphorylated. It appears, therefore, that the enzyme is highly specific for particular serine residues in many different proteins.

It seems unlikely that the acceptor serine-residue specificity can depend entirely upon the three-dimensional structure of the protein since such a wide variety of enzymes are affected. In many cases denaturation of a protein does increase its susceptibility to phosphorylation, and serine residues that were not phosphorylated in the native protein may become phosphorylated. It appears then that simple accessibility of the serine residues may be a factor and that some residues which would be phosphorylated are protected by the folding of the polypeptide chain. However, even in a denatured protein many of the serines are not phosphorylated, suggesting that the sequence of amino acids around the

Table 4.1. Some enzymes regulated by cyclic AMP-dependent phosphorylation

	Effect
Phosphorylase kinase	Activation
Glycogen synthetase	Inhibition
Phosphofructokinase 2/ fructose 6-phosphatase 2	Inhibition/activation
Pyruvate kinase	Inhibition
Acetyl CoA carboxylase	Inhibition
Triacylglycerol lipase	Activation
Inhibitor 1	Activation
Phenylalanine hydroxylase	
Troponin 1	
Cholesterol esterase	
Smooth muscle myosin light chain kinase	Inhibition

substrate serine residue may be important. Two experimental approaches have confirmed this. In the first, proteins were phosphorylated in their native state using ^{32}P-ATP. The proteins were then digested with trypsin and a tryptic peptide map was made. The phosphorylated serine-containing peptides could then be sequenced. The second approach was to test small synthetic peptides containing serine as substrates for cyclic AMP-dependent protein kinase. From these studies it appears that, when a serine residue is a substrate for cyclic AMP-dependent protein kinase, there is always an arginine residue between two and five amino acid residues on the N-terminal side of the serine. Several of the best natural substrates for the kinase contain two basic amino acids close to the target serine, one of which is arginine. This appears to be the major factor determining the specificity of the enzyme.

4C Structure and mechanism of action of the protein kinase

The inactive protein kinase is a tetramer made up of two types of subunit: two regulatory subunits (R) and two catalytic subunits (C). Each R unit contains two binding sites for cyclic AMP. At saturating cyclic AMP concentrations the enzyme dissociates to leave an R dimer and two free catalytic subunits:

$$R_2C_2 + 4\ cAMP \rightarrow R_2\ cAMP_4 + 2C.$$

As a consequence of the dissociation, the catalytic units become active. Yeast R unit will inhibit mammalian C unit in the absence of cyclic AMP, and mammalian R unit will similarly inhibit yeast C. The proteins therefore seem to have been highly conserved during evolution.

4C1 Cellular location

A major proportion of the cyclic AMP-dependent protein kinase activity is cytoplasmic. However, membrane fractions, in particular the plasma membrane, contain substantial amounts of activity. Membrane-bound cyclic AMP-dependent protein kinases are relatively poorly characterized, but it seems that membrane location is a function of the R subunit rather than the C subunit. Plasma membrane-bound cyclic AMP-dependent protein kinase in liver is modulated by insulin (see Chapter 10), and a plasma membrane location does raise the possibility of direct modulation of the enzyme activity by hormone–receptor complexes.

4C2 Type I and type II enzyme

In most mammalian tissues, two distinct fractions of protein kinase can be isolated which are usually referred to as type I and type II. They are different in net charge and can therefore be separated by ion-exchange chromatography; they also differ slightly in size. The catalytic units of the type I and type II enzyme appear to be identical, or at least very similar. They have a molecular weight of 40 000. There is no difference in physical or chemical properties between C subunit isolated from purified type I enzyme, and C subunit isolated from purified type II enzyme. The catalytic properties also appear to be identical, with the same K_m for ATP and the same substrate specificity. Monoclonal antibodies to the type II C subunit interact equally effectively with the type I C subunit. Furthermore, C subunit obtained from purified type I enzyme interacts with type II R subunit, just as well as C subunit derived from purified type II enzyme, and vice versa.

The distinction between type I and type II enzyme resides in the R subunits which have molecular weights of 49 000 for type I, and 55 000 for type II, and are clearly different. A possibility to be considered is that the smaller RI is derived from RII by proteolysis. This has been excluded by an examination of the tryptic peptide maps for the two types, which are markedly different. Also, antibodies against one R type do not recognize the other.

The two types of R subunit also have different functional properties, and show differences in response to cyclic AMP. The type I form has a higher affinity than the type II form for cyclic AMP. Half-maximal activation of type I takes place at 0.01 to 0.03 µmol/l, while the type II form requires 0.09 to 0.15 µmol/l. These figures, which were determined with pure enzyme in dilute solution, need to be treated with caution. The precise assay conditions are important, factors such as the protein substrate concentration, the concentration of the enzyme itself, and the ATP concentration, having substantial effects. The type I form, but not the type II form, is affected by the concentration of $MgATP^{2-}$, increasing levels leading to an increase in the concentration of cyclic AMP required for half-maximal activation. The effects of $MgATP^{2-}$ appear to be the result of non-covalent binding rather than phosphorylation. The type II enzyme is not affected by $MgATP^{2-}$ binding. Recently, the binding of both types of R subunit has been measured for a number of analogues of cyclic AMP. Differences in binding affinity suggest that there are differences between the two cyclic AMP binding sites on the same subunit, and also differences between the binding sites on the RI and RII subunits.

Both types of R subunit undergo phosphorylation. The RII subunit has two phosphorylation sites, one close to each cyclic AMP binding site. One of the sites is subject to phosphorylation by the C subunit by an intramolecular autophosphorylation mechanism. This phosphorylation reduces the rate of association of RII and C. The type II R subunit is also phosphorylated by glycogen synthetase kinase-3. This is interesting since glycogen synthetase kinase-3 is thought to be inhibited by insulin (see Chapter 14). Thus there appear to be mechanisms which may modulate the response of RII to cyclic AMP and, hence, the activation of the protein kinase. It should be stressed, however, that as yet no physiological role has been established for these effects. The type I R subunit is not autophosphorylated, but is phosphorylated by a cyclic GMP-dependent protein kinase. Again, the physiological significance of this, if any, is unknown.

It is not entirely clear whether the differences between type I and type II enzyme have any importance *in vivo*. However, it seems likely that the differences are of significance, since the proportions of the kinases vary markedly from one tissue to another in the same species. In the rat, for example, eighty per cent is in the type I form in the heart while almost all the enzyme in adipose tissue is in the type II form. Differences in the proportions of the two forms may reflect differences in sensitivity of the tissues to rises in cyclic AMP. If the phosphorylation reactions described above are significant in the regulation of protein kinase activity *in vivo*, the two different types may respond differently to modulation by factors other than cyclic AMP. In this case the sensitivity of the cyclic AMP response to other agonists may differ between cell types depending upon the proportion of type I and type II enzyme. However, at this stage there is insufficient evidence to do more than speculate.

4C3 Activity of the enzyme in the intact cell

It has been shown in a very wide range of different cell types that a rise in cyclic AMP in response to a hormone produces a rise in the activity of the cyclic AMP-dependent protein kinase. However, there is, apparently, a discrepancy between the basal protein kinase activity and the resting levels of cyclic AMP in the cell. The concentration of cyclic AMP in a resting cell ranges from 0.1–1.0 μmol/l, while K_a for cyclic AMP of the protein kinase is in the range 0.01–0.15 μmol/l. At first sight, it appears that there is enough cyclic AMP in the resting cell to substantially activate protein kinase. However, the K_a value was determined using pure enzyme in dilute solution, and the concentration of protein kinase was much less than the K_a determined. In the cell, the concentration of protein kinase is much greater at about 0.2 μmol/l. Thus the concentration of enzyme is comparable with the concentration of cyclic AMP, and much of the cyclic AMP will be bound. Since only a small proportion of cyclic AMP is actually in solution, this has the effect of increasing the apparent K_a (Fig. 4.1), and if the K_a for cyclic AMP is determined at physiological concentrations of protein kinase, a figure of 1.5 μmol/l is obtained which is well in excess of the resting levels in most tissues.

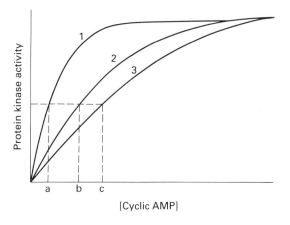

Fig. 4.1. Effect of increasing protein kinase concentration on the apparent K_a for cyclic AMP. As the concentration of protein kinase increases (curves 1–3), the proportion of total cyclic AMP present which is bound to R subunits increases. As a result, a higher total cyclic AMP is needed to generate a given free cyclic AMP concentration, and the apparent K_a increases (a–c).

4C4 Inhibitory protein

The resting activity of protein kinase is also affected by the existence of an inhibitory protein factor in many tissues that inhibits cyclic AMP-dependent protein kinase. The protein has a molecular weight of 11 000. It is heat stable and resistant to most conventional denaturing agents, which suggests that it has little three-dimensional structure in solution. It only affects the active dissociated C unit and seems to have no effect upon the process of dissociation.

The function of the inhibitor protein is not clear, but one plausible explanation has been suggested. In some tissues, notably the heart, there is enough inhibitor to block the activity of 20% of the protein kinase. Thus as the level of cyclic AMP rises to the point where 20% of the catalytic subunits dissociate, there will be little or no rise in enzyme activity. Further increases in cyclic AMP concentration will produce a sharp rise in activity. Other tissues contain much lower concentrations of inhibitor protein. The inhibitor protein offers the possibility of setting the range of cyclic AMP concentration over which the protein kinase will respond (Fig. 4.2).

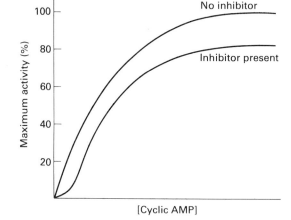

Fig. 4.2. Effect of inhibitor protein on the activation of protein kinase by cyclic AMP. The figure shows the effect of the presence of enough inhibitor to account for 20% of the total kinase activity. Thus no activation is seen until 20% of the kinase is dissociated, and the maximum possible activity, when all the kinase is dissociated is reduced by 20%.

4D Conclusions

The regulation of the production of cylic AMP is very sophisticated. The protein phosphorylation reaction can also be modulated by multiple effectors acting upon the protein substrates (see Chapter 1). On the basis of the evidence currently available, it seems that the control of the cyclic AMP dependent-protein kinase itself is relatively simple. It seems likely that the kinase activity in a particular cell type has a more or less constant sensitivity to changes in cyclic AMP concentration. The role of the different R subunits in the type I and type II enzyme, and of the inhibitor protein, seems to be to allow variations in sensitivity to changes in cyclic AMP between different cell types, rather than to allow the sensitivity to be modulated in a single cell type. However, as described in Section 4C1, there is some preliminary evidence to suggest that more complex interactions may take place.

Suggestions for further reading

Beavo J.A., Bechtel P.J. & Krebs E.G. (1975) Mechanisms of control of cyclic AMP dependent protein kinase. *Advances In Cyclic Nucleotide Research*, **5**, 241–252.

Corbin J.D., Keely S.L., Soderling T.R. & Park C.R. (1975) Hormonal regulation of cyclic AMP dependent protein kinase. *Advances In Cyclic Nucleotide Research*, **5**, 265–281.

Corbin J.D. & Lincoln T.M. (1978) Comparison of cyclic AMP dependent and cyclic GMP dependent protein kinases. *Advances In Cyclic Nucleotide Research*, **9**.

Rosen O.M., Erlichman J. & Rubin C.S. (1975) Molecular structure and characterisation of bovine heart protein kinase. *Advances In Cyclic Nucleotide Research*, **5**, 253–264.

Chapter 5 Ca^{2+} and cellular regulation

5A Introduction

Calcium ions have been implicated in many regulatory processes in mammalian cells. Changes in Ca^{2+} levels are involved in muscle contraction, the control of enzyme activity, and the exocytosis of proteins and neurotransmitters from cells. Calcium has also been proposed as a second messenger for the action of hormones and neurotransmitters. It has been suggested that changes in the concentration of free Ca^{2+} in the cytoplasm are important in the mechanism of action of catecholamines, vasopressin, insulin and many other hormones. It is also thought that Ca^{2+} ions are important in the response of cells to other types of external stimulation, in particular the response to factors which control cell growth and proliferation. In this chapter the control of Ca^{2+} levels in the resting cell will be considered. The role of hormones in altering cellular Ca^{2+} concentration, and the mechanisms by which changes in Ca^{2+} levels affect the metabolism of the cell, will be considered in Chapters 6, 7 and 8.

5B Control of cellular Ca^{2+} levels

Calcium is, in many respects, less satisfactory as a regulator than cyclic AMP. The concentration of cyclic AMP in the cytoplasm is regulated by the interaction of two enzyme activities: one to synthesize and one to break down the second messenger. Both can be subject to control. Cyclic AMP does not have any part to play in the mainstream of metabolism and simply functions as a regulator (see Chapters 3 and 4). Ca^{2+} ions, however, are a major component of the extracellular fluid; serum contains 2.6 mmol/l total Ca^{2+} and about 1.3 mmol/l free Ca^{2+}. Thus there are substantial amounts of calcium in the environment of the cell, and the control of the free intracellular Ca^{2+} concentration requires a different type of mechanism to that used for the control of cyclic AMP levels.

The regulation of intracellular Ca^{2+} levels is very important for the normal function of the cell. Table 5.1 shows the concentrations of the major inorganic cations in serum and in the cytoplasm. While the total concentration of calcium in the cell is similar to that in serum, the free Ca^{2+} concentration in cytoplasm is lower than the free Ca^{2+} in serum by a factor of about 10^4. The rest of the calcium in the cell is sequestered in organelles or bound to protein or other charged molecules. Thus the gradient of Ca^{2+} concentration across the plasma membrane is much greater than for any other ion.

There are three main barriers across which calcium may enter or

Table 5.1. Concentrations of cations in serum and cytoplasm

	Serum		Cytoplasm	
	Total	Free	Total	Free
K^+	5 mM	5 mM	155 mM	155 mM
Na^+	145 mM	145 mM	12 mM	12 mM
Mg^{2+}	2 mM	2 mM	20 mM	3 mM
Ca^{2+}	2.8 mM	1.3 mM	100 μM	0.1 μM

leave the cytoplasm: the plasma membrane, the inner mitochondrial membrane, and the membrane of the endoplasmic reticulum. All three processes take place in most cell types but their relative importance can vary depending upon the specialized function of the cell.

5C Calcium transport across the plasma membrane

5C1 Influx across the plasma membrane

There is a very large concentration gradient for Ca^{2+} between the extracellular fluid and the cytoplasm. The Ca^{2+} electrochemical gradient is even larger, so it is not surprising that there is a continuous influx of Ca^{2+} into the cell. In excitable cells such as muscle, calcium influx is clearly important in the regulation of cell function. While the onset of muscle contraction is mediated by a release of Ca^{2+} ions from the sarcoplasmic reticulum, the maintenance of contraction requires the presence of extracellular calcium. Ca^{2+} ions enter excitable tissues via two pathways. There is an entry which coincides with the onset of the action potential which probably takes place via the Na^+ channel. The second channel is activated by depolarization of the plasma membrane and is known as the voltage dependent channel. This is specific for Ca^{2+} and is blocked by the drug verapamil and a number of related drugs which can be shown to have specific binding sites in the plasma membrane.

In non-excitable cells such as liver or fat cells, the situation is less clear. In some non-excitable cell types, drugs directed against the voltage dependent channel affect Ca^{2+} influx, and specific binding sites for the drugs can be found on the surface of the cell. In other cell types the drugs have no effect. The existence of non-voltage dependent Ca^{2+} channels which may be regulated by hormones has also been proposed. It seems likely that non-excitable cells do contain specific proteins responsible for Ca^{2+} influx, but as yet these are relatively poorly characterized.

5C2 Calcium efflux across the plasma membrane

There are two distinct mechanisms for pumping Ca^{2+} ions from the cytoplasm to the extracellular space across the plasma membrane.

1 *The Ca^{2+}-ATPase.* This mechanism has been extensively studied in the erythrocyte plasma membrane. Erythrocytes are the ideal system in which to study the plasma membrane Ca^{2+}-ATPase. The activity of other ATPases is low, and a particular advantage is the lack of mitochondria and, accordingly, the mitochondrial ATPase. Plasma membrane ghosts can be prepared making it possible to study the sidedness or topography of the system. The Ca^{2+}-ATPase was found to be an integral membrane protein with a molecular weight of 130 000. In common with all other ATPases, it makes use of $MgATP^{2-}$ as substrate and, as a result, is sometimes referred to as the Ca^{2+}/Mg^{2+}-dependent ATPase. ATP binds to a site on the inner surface of the plasma membrane, and the K_m for $MgATP^{2-}$ is about 5×10^{-5} mol/l. This is about a hundredfold lower than the normal cytoplasmic ATP concentration so that the enzyme is never limited by the availability of ATP as substrate.

In other cell types it has been shown that Ca^{2+} extrusion is dependent upon metabolic energy, and at least partly independent of the gradient of Na^+ ions across the plasma membrane. However, in cells other than erythrocytes it is difficult to demonstrate the existence of a Ca^{2+}-dependent ATPase in broken cell preparations. This is normally done by measuring the total ATPase activity in the presence and absence of the Ca^{2+} chelator, EGTA, and taking the difference. This is often a small proportion of the total since the sum of cellular ATPase activity is high. Furthermore, conditions are chosen to minimize the bulk ATPase activity, but in many cases such conditions are probably sub-optimal for the Ca^{2+}-ATPase. In spite of the difficulties of measuring Ca^{2+}-dependent ATPase activity, it is generally assumed that an ATP-driven Ca^{2+} pump makes a major contribution to the efflux of Ca^{2+} from the cell, and there is a considerable amount of circumstantial evidence for this. Recently the Ca^{2+}-ATPase has been isolated from liver plasma membranes providing direct evidence for its existence. Obviously alterations in the activity of such a pump would provide a possible mechanism for changing the cytoplasmic free Ca^{2+} concentration. There has been a lot of interest in this possibility but so far only one report of a specific regulation of the Ca^{2+}-ATPase, insulin having been reported to inhibit the Ca^{2+}-ATPase in fat cell plasma membranes. In erythrocytes, the activity of the Ca^{2+}-ATPase is affected by cytoplasmic free Ca^{2+} levels acting via the regulator protein, calmodulin. Both these points will be considered later.

2 *The Na^+/Ca^{2+} exchanger.* There is good evidence that the plasma membrane of many cell types contains a carrier which catalyses the exchange of Na^+ ions for Ca^{2+} ions. It was first demonstrated in the giant axon of squid. The cells were preloaded with the radioactive isotope ^{45}Ca, and the rate of release of ^{45}Ca from the cell measured. It was shown that the removal of Na^+ from the external medium led to a reduction in the rate of ^{45}Ca efflux from the cell. This led to the suggestion that there was a Na^+/Ca^{2+} exchange carrier.

The squid giant axon is such a large cell that it is possible to manipulate the contents of the cytoplasm by microinjection, or to measure the concentration of ions in the axoplasm (cytoplasm) by the insertion of microelectrodes. The axoplasm can be directly replaced with a defined medium making it possible to examine the effect of a gradient of one ion upon the transport of another. At the same time, ATP can be removed so that no ATPase-driven pumps can function. The effect of varying the Na^+ gradient on the transport of Ca^{2+} was examined. The results suggested that there was a carrier which exchanged three Na^+ ions for one Ca^{2+} ion. The carrier was also found to be able to exchange Ca^{2+} for Ca^{2+} with a 1:1 stoichiometry. The exchange of three Na^+ for one Ca^{2+} will lead to the transfer of one positive charge across the cell membrane. The exchange will therefore be affected by the membrane potential. Since the plasma membrane potential is net negative inside, hyperpolarization will favour Na^+-dependent Ca^{2+} efflux.

The driving force for Ca^{2+} efflux, resulting from the combined effects of the Na^+ gradient across the plasma membrane and the membrane potential, can be calculated. When this was done for the squid axon it was found that, at equilibrium, the known Na^+ gradient would support the known Ca^{2+} gradient across the membrane if there was a net inward transport of between one and two positive charges for each Ca^{2+} transported out. Thus the Na^+/Ca^{2+} exchanger is close to being able to account for the Ca^{2+} gradient. More recent calculations suggest that in the squid giant axon, the Na^+/Ca^{2+} exchanger and the Ca^{2+}-ATPase make a roughly equal contribution to the maintenance of the Ca^{2+} gradient.

Na^+/Ca^{2+} exchange has been demonstrated in many mammalian tissues including brain, pancreas, adrenal medulla, muscle and fat cell. In cells where it is significant in maintaining the Ca^{2+} gradient across the plasma membrane, the exchange carrier could be used to alter Ca^{2+}

efflux and hence the cytoplasmic free Ca^{2+} concentration. This might be achieved by a direct effect on the carrier, by changes in the intracellular Na^+ concentration, or, since the carrier is electrogenic, by alterations in the membrane potential. Many hormones are known to affect the plasma membrane potential of target cells.

At present it is not clear in most cell types which of the two available transport systems is primarily responsible for the transport of Ca^{2+} out of the cell. In erythrocytes, the Ca^{2+}-ATPase is clearly the major mechanism. In nerve cells, both carriers appear to be important and it seems likely that this is the case in many other cell types as well.

5D The transport of Ca^{2+} by the endoplasmic reticulum

Interest in a possible role for the endoplasmic reticulum of non-muscle cells in controlling cytoplasmic free Ca^{2+} levels is relatively recent. The endoplasmic reticulum of brain, liver, kidney, fat cells, salivary gland, and platelets have all been shown to have the capacity to accumulate Ca^{2+}, so it seems likely that this is a general phenomenon. In all the systems examined so far, the source of energy for the accumulation is $MgATP^{2-}$. Reported values of the affinity for both Ca^{2+} and ATP in isolated endoplasmic reticulum preparations vary widely depending upon the tissue. Values for the K_m for Ca^{2+} vary from 4.6 µmol/l in liver, to as much as 100 µmol/l in some other tissues, and values for the K_m for ATP vary between 20 µmol/l and 1.8 mmol/l. The uptake of Ca^{2+} by the endoplasmic reticulum can be distinguished from plasma membrane transport in that it is inhibited by Na^+ ions. This helps to exclude the possibility of Ca^{2+} transport by contaminating fragments of plasma membrane in endoplasmic reticulum preparations. At first sight it seems unlikely that the endoplasmic reticulum can contribute significantly to the control of cytoplasmic free Ca^{2+} levels. The K_m for Ca^{2+} appears to be at least 5 µmol/l against a likely free Ca^{2+} concentration in the cytoplasm of 0.1 µmol/l in the resting cell, and the total capacity is comparatively small. However, studies on the endoplasmic reticulum are at a relatively early stage and it is technically difficult to purify endoplasmic reticulum fractions.

Intracellular pools of Ca^{2+} have been measured by calcium flux studies in liver cells in which the plasma membrane had been made freely permeable to Ca^{2+} by the addition of the detergent digitonin. Two pools were identified. The larger pool responded to inhibitors of mitochondrial electron transport and oxidative phosphorylation in a manner which suggested that it represented the mitochondria. The smaller pool had a higher affinity for Ca^{2+} ions and was suggested to represent the endoplasmic reticulum. There have been few studies on the release of Ca^{2+} ions from the endoplasmic reticulum or of possible mechanisms for controlling release. However, one recent report does suggest that calcium-mobilizing hormones cause release of Ca^{2+} ions from the endoplasmic reticulum and a mechanism is proposed (see Section 9H4). The possibility that calcium transport by the endoplasmic reticulum membrane may be important in the regulation of cytoplasmic free Ca^{2+} levels cannot be excluded.

5E The sarcoplasmic reticulum

In muscle cells, the specialized endoplasmic reticulum, or sarcoplasmic reticulum, plays a key role in the control of Ca^{2+} levels in the cell. Calcium is the key regulatory factor initiating muscle contraction. Fast muscle contains a great deal of sarcoplasmic reticulum, while slow muscle and heart contain much less.

5E1 Calcium uptake by the sarcoplasmic reticulum

The membrane of the sarcoplasmic reticulum contains a Ca^{2+}-dependent ATPase which is responsible for pumping Ca^{2+} into the reticulum lumen. It is a protein with a molecular weight of 100 000 and it accounts for as much as eighty per cent of the total membrane protein. Because of this very high abundance, the sarcoplasmic reticulum Ca^{2+}-ATPase was relatively easy to purify and it is well characterized. By way of comparison, the erythrocyte plasma membrane Ca^{2+}-ATPase accounts for about 0.2 per cent of the total membrane protein. Pure sarcoplasmic reticulum Ca^{2+}-ATPase can be reconstituted into phospholipid vesicles and will then pump Ca^{2+}.

The Ca^{2+}-ATPase of sarcoplasmic reticulum appears to pump two Ca^{2+} ions for each ATP hydrolysed. This is suggested by a number of observations. Comparison of rates of Ca^{2+} influx and ATP hydrolysis in artificial vesicles gives a stoichiometry of about 2, the enzyme has a Hill coefficient for Ca^{2+} of 1.8, implying at least two binding sites for Ca^{2+}, and two high-affinity Ca^{2+}-binding sites have been detected by direct binding studies. The concentration of Ca^{2+} for half-maximal velocity is about 0.5 µmol/l. The pump requires $MgATP^{2-}$, and the concentration needed for half-maximal velocity is about 5 µmol/l. The pump is thought to function by the following sequence of events which are summarized in Fig. 5.1.

1 Two Ca^{2+} ions and one ATP are bound on the cytoplasmic side of the membrane.

2 The terminal phosphate is transferred to an aspartate side chain on the enzyme. This is dependent on the presence of Ca^{2+}, and the equilibrium constant is close to one. A result of this is that it is possible to devise conditions *in vitro* when the pump will synthesize ATP from ADP and P_i.

3 The protein undergoes a conformational change leading to the translocation of the Ca^{2+} to the lumen surface of the plasma membrane.

4 Two Ca^{2+} ions are released into the lumen of the sarcoplasmic reticulum and one Mg^{2+} ion bound.

5 The enzyme is dephosphorylated, returning to the ground state and releasing a single Mg^{2+} into the cytoplasm. Since two Ca^{2+} ions are transported, a further two positive charges need to be transported from the extracellular space to the cytoplasm to make the process electrically neutral. It is thought that the other two positive charges are balanced by the transport of two K^+ ions.

Dephosphorylation is the rate-limiting step of the process. As a result, high levels of Ca^{2+} in the lumen will inhibit Ca^{2+} uptake by preventing Ca^{2+} release and dephosphorylation. The apparent Ca^{2+}

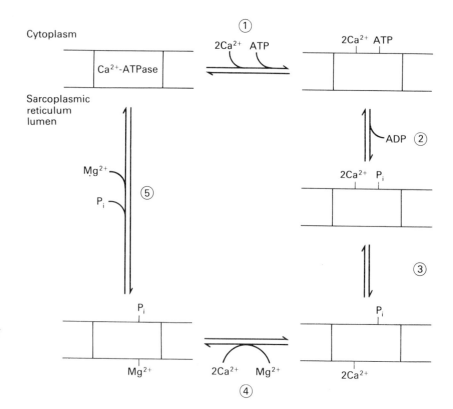

Fig. 5.1. Mechanism of the sarcoplasmic reticulum Ca^{2+}-ATPase. See text for description of numbered steps.

gradient across the sarcoplasmic reticulum membrane, if it is assumed that all the Ca^{2+} ions are free, is 10^3, implying an internal concentration of about 1 mmol/l. This is a high enough concentration to inhibit the calcium-ATPase, suggesting that some of the calcium must be bound to prevent inhibition of the pump. Some Ca^{2+} will be bound to low-affinity sites on the membrane, but the most important source of binding is the protein calsequestrin. This has a K_d for Ca^{2+} of 8×10^{-4} mol/l, and a very high specificity for Ca^{2+}. A second Ca^{2+}-binding protein with a rather higher affinity for Ca^{2+} has also been described. Thus the lumen of the sarcoplasmic reticulum has considerable buffering capacity for Ca^{2+} which will prevent the free Ca^{2+} concentration from rising to levels which will inhibit the Ca^{2+}-ATPase.

5E2 Ca^{2+} release from the sarcoplasmic reticulum

The mechanism of release of Ca^{2+} from the sarcoplasmic reticulum is much less well characterized than the mechanism of uptake. Vesicles of sarcoplasmic reticulum can be isolated and are capable of accumulating Ca^{2+}. The rate of release of Ca^{2+} from such preparations is about 1000 times slower than the rate of uptake. Rapid release of Ca^{2+} can take place by a reversal of the uptake process. However, this requires a low external Ca^{2+} concentration and high levels of ADP, P_i and Mg^{2+}. There is no evidence that the level of ADP rises before muscle contraction, as might be expected if release occurred by a reversal of the uptake process. It appears that release is caused by depolarization of the plasma membrane. The precise mechanism has not been established but two suggestions have been put forward. The first proposes that depolarization in some way directly promotes Ca^{2+} release. The second hypothesis suggests that depolarization of the plasma membrane leads to a small influx of Ca^{2+} through the voltage dependent Ca^{2+} channel and that this small increase in cytoplasmic Ca^{2+} triggers the release of Ca^{2+} from the sarcoplasmic reticulum.

5F Mitochondrial Ca^{2+} uptake

Respiration dependent accumulation of Ca^{2+} ions by mitochondria was first discovered in the 1960s. Mitochondria from all mammalian cell types have a high capacity to retain Ca^{2+} and will accumulate Ca^{2+} at low external free Ca^{2+} concentrations. It was not, at first, obvious how the energy of respiration was to be coupled to the accumulation of an ion across the inner mitochondrial membrane. An explanation was provided by the chemiosmotic hypothesis of Mitchell (1961) which proposed that the energy of respiration is used to pump H⁺ ions out of the mitochondria setting up an electrochemical gradient, or proton-motive force, $\Delta\mu H$, which has a value of about 230 mV. Of this, about 150–180 mV consists of a membrane potential, $\Delta\psi m$, and the rest consists of a pH gradient, ΔpHm. The proton-motive force is used to drive the synthesis of ATP from ADP and P_i. If the membrane is permeable to a cation, the proton-motive force can be used to accumulate the cation against a concentration gradient. The mitochondrial membrane does contain Ca^{2+} carriers.

Two other key observations were made. In the absence of anions which can enter the mitochondria, the accumulation of Ca^{2+} was strictly limited and two H⁺ ions were released from the mitochondria for each Ca^{2+} taken up (Fig. 5.2a). In the presence of an anion which can enter the mitochondria, e.g. acetate, glutamate, bicarbonate, β-hydroxybutyrate or phosphate, uptake was much more extensive and the strict stoichiometry between H⁺ release and Ca^{2+} was lost.

The effect of the presence of an anion which can enter the mitochondria can be explained by considering the thermodynamics of the system. The Ca^{2+} electrochemical potential is given by the expression

$$\Delta\mu Ca^{2+} = 2\Delta\psi m - 60 \log \frac{[Ca^{2+}] \text{ mitochondrial}}{[Ca^{2+}] \text{ medium}}$$

where $\Delta\psi m$ is the mitochondrial membrane potential. This assumes

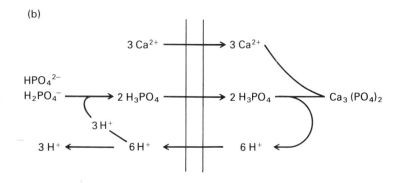

Fig. 5.2. Mitochondrial Ca^{2+} uptake. (a) In the absence of permaent anions. A strict stoichiometry of 1 Ca^{2+} for 2 H⁺ is maintained, and uptake is limited by the thermodynamics of the system. (b) In the presence of phosphate. The constraint is removed by the presence of a permaent anion, under physiological conditions usually phosphate. See text for full explanation.

that uptake of Ca^{2+} is electrogenic involving a net uptake of charge with no balancing ion movements. Uptake of Ca^{2+} will therefore rapidly discharge the membrane potential and the mitochondria will respond by increased respiration and the extrusion of H^+ ions. This leads to an increase in the intramitochondrial pH, and hence the pH gradient, ΔpHm. The total value of $\Delta \mu H^+$ is limited to 230 mV by the thermodynamics of the redox reactions in the respiratory chain, so that an increase in the pH gradient results in a reduction in the membrane potential limiting the possible accumulation of Ca^{2+} ions. If a negative ion is present to enter the mitochondria, either together with H^+ or in exchange for OH^-, the build up of the pH gradient is prevented and accumulation of Ca^{2+} can be much more extensive. Under physiological conditions, phosphate is likely to be the anion transported with Ca^{2+}. Phosphate has the advantage that the insoluble salt, $Ca_3(PO_4)_2$, is formed preventing an increase in the osmolarity of the mitochondria which would result in an influx of water causing the mitochondria to swell. The pK for the second dissociation of H^+ from phosphate is about 7.0. Since the cytoplasmic pH is about 7.3, the transport of one phosphate into the mitochondria in effect transports in 1.5 H^+ ions. Under these conditions there is a net appearance of one H^+ in the medium for each Ca^{2+} accumulated, but no appearance of OH^- in the mitochondria (Fig. 5.2b). In the intact cell, the H^+ released into the cytoplasm will be buffered.

5F1 Mechanism of the uptake carrier

It appears that uptake of Ca^{2+} requires the presence of an anion, and that under physiological conditions the anion is phosphate. The two ions are transported on separate carriers, phosphate in exchange for hydroxyl ions, and Ca^{2+}, by a direct uniport mechanism with no linked ion movements, on the uptake carrier. It has been suggested that the carrier co-transports Ca^{2+} and HPO_4^{2-} by a symport mechanism, but this is very unlikely. Uptake is not completely dependent upon the presence of an anion (see above), and other anions can substitute for phosphate in increasing the extent of uptake.

5F2 Factors affecting the rate of Ca^{2+} uptake

The rate of Ca^{2+} uptake into mitochondria is affected by external Ca^{2+} concentration, pH, temperature and polycations. Transport appears to be unidirectional in the presence of a sufficiently large membrane potential, and large variations in the intramitochondrial Ca^{2+} concentration have little effect. The rate of uptake is very dependent on the external Ca^{2+} concentration with a K_m for Ca^{2+} of about 10^{-6} mol/l. The V_{max} appears to be limited by the ability of the mitochondria to maintain a membrane potential by respiration. This conclusion is supported by the observation that faster rates of uptake can be achieved by an artificially generated membrane potential, formed using a K^+ ion gradient and the ionophore valinomycin, to render the mitochondria permeable to K^+.

The concentration of Ca^{2+} needed for a given rate of uptake is increased in the presence of Mg^{2+} ions, and increasing the extramitochondrial pH from 7 to 8 markedly increases the rate of uptake. It is not certain whether this is a direct effect, or if it results from a reduction in the pH gradient across the mitochondrial membrane causing an increase in the membrane potential (see above). The rate of uptake is relatively insensitive to temperature when driven by an artificially generated K^+ gradient, rather than by a gradient generated by electron transport. The carrier is also difficult to saturate at high Ca^{2+} levels and the activation energy of the process is very low. These two observations suggest a channel mechanism rather than a pump. Uptake is specifically inhibited by the dye ruthenium red, which reacts with glycoproteins, so the carrier may be a glycoprotein.

5G Ca^{2+} efflux from the mitochondria

Any mechanism allowing efflux of Ca^{2+} from the mitochondria will operate against the membrane potential. As a result there must be a linked ion flow on the same carrier to maintain electrical neutrality and an electrogenic uniport can be ruled out. There are a number of lines of evidence to support the existence of an efflux carrier.

1 The use of the specific inhibitor of the uptake uniporter, ruthenium red, causes net loss of Ca^{2+} from the mitochondria.

2 At low mitochondrial membrane potentials, the distribution of Ca^{2+} across the mitochondrial membrane is close to equilibrium with the membrane potential. At higher potentials it is considerably lower than expected, suggesting a coupled efflux pathway.

3 If the electrogenic influx uniporter were the only carrier present we would expect a gradient of Ca^{2+} of $10^5 - 10^6$ across the membrane. The cytoplasmic free Ca^{2+} concentration varies between 10^{-7} mol/l and 10^{-6} mol/l in most cell types. This implies that the mitochondria contain molar concentrations of Ca^{2+} orders of magnitude above the solubility product for calcium phosphate. An efflux carrier removes the constraint that the Ca^{2+} gradient must be in equilibrium with the membrane

Ca^{2+} and cellular regulation

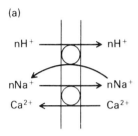

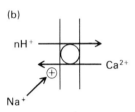

Fig. 5.3. Possible mechanisms for Na^+-dependent Ca^{2+} efflux from mitochondria. (a) Exchange. Ca^{2+} leaves the mitochondria in exchange for Na^+ which in turn leaves in exchange for H^+. (b) Direct activation. Ca^{2+} exchanges directly for H^+ and Na^+ activates the carrier, binding on the outside.

potential. Two types of efflux carrier have been identified: Na^+ dependent and Na^+ independent.

5G1 Na^+-dependent efflux

Na^+-dependent carriers have been found in heart, skeletal muscle, parotid, kidney, and brown fat. In mitochondria from all these sources, Ca^{2+} efflux is activated by Na^+ at mmol/l concentrations. There are two possible mechanisms: a Na^+/Ca^{2+} exchange carrier, or a Na^+-activated exchange of Ca^{2+} for H^+ (Fig. 5.3). It is difficult to distinguish between these two possibilities since mitochondria contain a carrier which is able to exchange Na^+ and H^+. This carrier is more active than the Ca^{2+} efflux pathway so that either mechanism can produce the same end result: an exchange of Ca^{2+} for H^+ (Fig. 5.3). However, the relationship between the rate of efflux of Ca^{2+} and the Na^+ concentration is sigmoid. This suggests that more than one Na^+ is involved per Ca^{2+} released, and led to the proposal that there is a carrier exchanging three Na^+ for one Ca^{2+}.

5G2 Na^+-independent efflux

Na^+-independent efflux has been studied most extensively in liver mitochondria. The rate of efflux depends upon the permeant anion present. Efflux rates are greater if acetate is present than if phosphate is present. In the presence of acetate the intramitochondrial free Ca^{2+} concentration will be higher than in the presence of phosphate, since calcium acetate does not form an insoluble complex. Therefore, it seems likely that the rate of the efflux depends upon the intramitochondrial free Ca^{2+} concentration. This applies to Na^+-dependent carrier mechanisms as well. In the presence of excess P_i, the formation of $Ca_3(PO_4)_2$ (Fig. 5.2b) probably maintains the free Ca^{2+} level below the level needed to saturate the carrier. This suggestion is supported by the observation that increasing the concentration of P_i in the external medium leads to a reduction in the rate of Ca^{2+} efflux. Under physiological conditions the mitochondria are exposed to a constant phosphate concentration in the cytoplasm which is well in excess of the free Ca^{2+} concentration. This has the effect of buffering the intramitochondrial free Ca^{2+} at a constant level which is largely independent of the total mitochondrial Ca^{2+}.

As in Na^+-dependent systems, Na^+-independent Ca^{2+} efflux requires linked ion movements to allow it to operate against the membrane potential, and must ultimately exchange for H^+ ions. This need not be a direct exchange, but a Ca^{2+}/H^+ exchange carrier seems to be the most likely mechanism.

5H Calcium cycling or buffering

In a normally functioning intact cell, the mitochondria will always have a substantial membrane potential and a substantial pH gradient. Therefore, under physiological conditions, both the Ca^{2+} influx carrier driven by the membrane potential, and the efflux carrier driven by the pH gradient, are unidirectional and irreversible. The use of separate unidirectional carriers for Ca^{2+} efflux and influx means that the mitochondria can maintain a constant external free Ca^{2+} over a wide range of internal total calcium. By analogy with chemical buffers, this is often referred to as mitochondrial calcium buffering. It results from the following properties of the system.

1 At a constant external phosphate concentration, the rate of the efflux carrier does not vary over a wide range of intramitochondrial calcium concentrations.

2 The rate of the influx carrier is highly dependent upon the external free Ca^{2+} concentration.

Therefore, mitochondria will accumulate Ca^{2+} from the external medium until the external free Ca^{2+} is sufficiently low to give an uptake rate equal to the constant efflux rate. The mitochondria will maintain the free external Ca^{2+} at this concentration. If the external concentration is reduced below this point, then there will be net efflux of Ca^{2+} until the external Ca^{2+} concentration is restored to the same steady state concentration. Thus the mitochondria function as a Ca^{2+} buffer.

With isolated mitochondria, the process is very simple to study using a Ca^{2+}-sensitive electrode which will give a direct read-out of the external free Ca^{2+} on a chart recorder. Addition of Ca^{2+} can be shown to result in Ca^{2+} uptake, while addition of the weak Ca^{2+} chelating agent, NTA, causes Ca^{2+} release (Fig. 5.4). The maximum rate of the uptake carrier is considerably greater than the maxium rate of the efflux carrier, with the result that the response to increased external Ca^{2+} is more rapid than the response to decreased external Ca^{2+}. The precise buffering point depends upon the relative rates of the two carriers. Values of between 0.3 µmol/l and 3 µmol/l have been reported in mitochondria from a wide variety of sources.

The Ca^{2+} buffering concentration will be affected by any factor which alters the rate of either the influx carrier or the efflux carrier. An increase in the rate of the influx carrier would lower the steady state

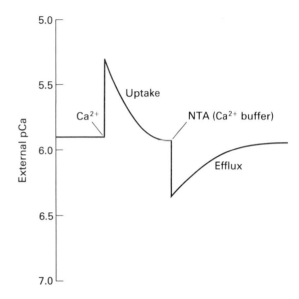

Fig. 5.4. Ca^{2+} buffering by isolated mitochondria. External Ca^{2+} concentration is measured continuously using a Ca^{2+} electrode. pCa^{2+} is the negative log of the Ca^{2+} concentration, analogous to pH. i.e. pCa^{2+} 6 corresponds to 10^{-6} mol/l Ca^{2+}.

buffered concentration, while an increase in the activity of the efflux carrier would raise the steady state concentration. Thus in mitochondria where the efflux carrier is Na^+-sensitive, increasing concentrations of Na^+ raise the buffering point. Mg^{2+} ions inhibit the uptake carrier and will therefore also raise the Ca^{2+} concentration at which the mitochondria will buffer. Increasing extramitochondrial phosphate reduces the rate of efflux and lowers the concentration at which the mitochondria buffer Ca^{2+}. Under physiological conditions, only Na^+ is likely to change so as to affect the buffering point.

Ca^{2+} buffering experiments are of necessity performed on isolated mitochondria or, at best, with whole cells where the plasma membrane has been made freely permeable to Ca^{2+}. Estimates of the Ca^{2+} buffering point *in vivo* should therefore be treated with caution. However, data from a number of sources suggests that mitochondria *in vivo* buffer Ca^{2+} at about 1 μmol/l.

The existence of independent influx and efflux pathways offers two important physiological advantages.

1 The mitochondria can respond to changes in the cytoplasmic Ca^{2+} concentration without changes in the membrane potential or the pH gradient. A single carrier system would require changes in these parameters which might interfere with ATP synthesis.

2 Relatively small changes in the activity of one of the carriers can cause relatively large changes in the net flux.

5J The intramitochondrial free Ca^{2+} concentration

The intramitochondrial free Ca^{2+} is influenced by the buffering effect of the formation of $Ca_3(PO_4)_2$. As a result, the rate of the efflux carrier is constant. However, this will only be the case at a constant phosphate concentration and an external Ca^{2+} concentration equal to, or in excess of, the Ca^{2+} buffering point. If the steady state extramitochondrial free Ca^{2+} is maintained below the Ca^{2+} buffering point, there will be a steady net loss of calcium from the mitochondria and eventually all the $Ca_3(PO_4)_2$ will dissociate. Under these conditions a change in extramitochondrial Ca^{2+} concentration will cause a change in mitochondrial Ca^{2+} concentration. If the extramitochondrial Ca^{2+} concentration is constant and below the buffering point, then changes in the rates of the carriers will affect the intramitochondrial free Ca^{2+} concentration.

The situation is further complicated by the fact that the dissociation of $Ca_3(PO_4)_2$ is relatively slow although the capacity of the $Ca_3(PO_4)_2$ to buffer Ca^{2+} (in other words, the amount of calcium present as calcium phosphate) is very large relative to other available sources of intramitochondrial calcium. Thus under steady state conditions the intramitochondrial $Ca_3(PO_4)_2$ may control the intramitochondrial free Ca^{2+}. However, if the intramitochondrial free Ca^{2+} changes as a result of a change in the rate of influx or efflux, the $Ca_3(PO_4)_2$ pool may be relatively slow to respond.

5K The control of cytoplasmic Ca^{2+} levels in the resting cell

The term 'resting cell' is taken to mean a cell in a normal physiological environment in the absence of hormones or any other external stimuli. Obviously this is not a situation which ever arises *in vivo* but it can be achieved, or at least a reasonable approximation can be achieved, *in vitro*.

Figure 5.5 shows the Ca^{2+} transport mechanisms which are available to a mammalian cell which has no specialized requirement for Ca^{2+} such as exists in muscle. There are three membrane barriers across which Ca^{2+} is transported: the plasma membrane, the mitochondrial membrane and the endoplasmic reticulum membrane. The capacity of the endoplasmic reticulum to accumulate Ca^{2+} is small in relation to the extracellular or intramitochondrial calcium pools. Transport across either the plasma membrane or the mitochondrial inner membrane, or both, probably provide the major mechanisms for the maintenance of a low concentration of Ca^{2+} in the cytoplasm. If this is the case, two situations can be envisaged in the resting cell.

Ca^{2+} and cellular regulation

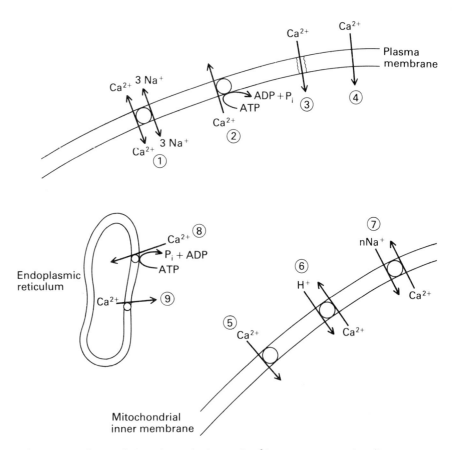

Fig. 5.5. Pathways for uptake and release of Ca^{2+}. Carriers: (1) Na^+/Ca^{2+} exchanger, (2) plasma membrane Ca^{2+}-ATPase, (3) voltage dependent Ca^{2+} channels, (4) other entry mechanisms, (5) mitochondrial uptake carrier, (6) mitochondrial efflux carrier, (7) Na^+-dependent efflux carrier, (8) endoplasmic reticulum Ca^{2+}-ATPase, (9) endoplasmic reticulum efflux carrier

In the first case the cytoplasmic free Ca^{2+} concentration is lower than the concentration at which the mitochondria buffer Ca^{2+}. If this is the case, the cytoplasmic free Ca^{2+} concentration is being controlled by the plasma membrane carriers, either the Ca^{2+}-ATPase or the Na^+/Ca^{2+} exchanger, or both. An explanation is needed for the failure of the mitochondria to release Ca^{2+} and so raise the cytoplasmic free Ca^{2+} concentration to the mitochondrial buffering point. There are two possibilities. Either the mitochondria may be totally depleted of calcium, or the mitochondria may contain calcium but the rate at which it can be mobilized as free Ca^{2+} and released is slower than the rate of net efflux across the plasma membrane. This would be the case if the capacity of the efflux carrier is slower than the rate of the carriers pumping Ca^{2+}

across the plasma membrane. Alternatively, the rate of Ca^{2+} release may be limited by the rate of mobilization of Ca^{2+} inside the mitochondria from $Ca_3(PO_4)_2$ and other bound sources. Mitochondrial calcium levels are difficult to determine accurately in intact whole cells, but it is clear that the mitochondria do contain substantial amounts of calcium under most conditions. If the cytoplasmic free Ca^{2+} is below the buffering point of the mitochondria, then it seems likely that Ca^{2+} release from the mitochondria is limited by the rate of efflux rather than by total depletion.

The second possible situation for the resting cell is that the cytoplasmic Ca^{2+} concentration is at the buffering point of the mitochondria; that is, the cytoplasmic free Ca^{2+} concentration is determined by the mitochondrial Ca^{2+} transporters. This implies that there is a continuous net influx of calcium into the cell that is accumulated by the mitochondria. The capacity of the mitochondria to accumulate Ca^{2+}, while substantial is finite, so this seems unlikely. Alternatively, we could postulate that the plasma membrane carriers maintain the cytoplasmic Ca^{2+} at a concentration which happens to be exactly the same as the mitochondrial buffering point, which also seems unlikely.

The hypothesis that the free cytoplasmic Ca^{2+} concentration is below the mitochondrial buffering point appears to be the more attractive. Measurements of cytoplasmic free Ca^{2+} concentration in resting cells give values in the region of 0.1 µmol/l. The Ca^{2+} buffering point of mitochondria is thought to be 1 µmol/l. The direct evidence therefore also favours the view that resting Ca^{2+} levels in the cytoplasm are below the mitochondrial buffering concentration. Finally, in considering these arguments it should be remembered that the concept of a resting cell is rather artificial. Cells *in vivo* will be responding to many different extracellular agents all the time.

Suggestions for further reading

Borle A.B. (1981) Control, modulation and regulation of cell calcium. *Rev Physiol Biochem Pharmacol*, **90**, 64–153.
Carafoli E. (Ed) (1982) *Membrane Transport of Calcium*. Academic Press.
Nicholls D.G. (1982) *Bioenergetics*. Academic Press.
Nicholls D.G. & Akerman K. (1982) Mitochondrial calcium transport. *Biochim Biophys Acta*, **683**, 57–88.

Chapter 6 Modulation of cytoplasmic Ca^{2+} by hormones

6A Introduction

Changes in cellular free Ca^{2+} levels play a major role in the response of mammalian cells to external stimuli. Ca^{2+} ions have been proposed as an intracellular signal for the action of many hormones, neurotransmitters and growth factors. The other major intracellular signal for rapid responses to external effectors is cyclic AMP. As described in Chapters 3 and 4, the mechanisms by which the cytoplasmic cyclic AMP concentration is controlled, and its mechanism of action, are relatively well understood. The role of Ca^{2+} is less well characterized. This reflects the fact that widespread interest in the role of Ca^{2+} as a regulator in non-excitable cells is more recent. Also, the problem is both technically more difficult and theoretically more complicated than studies on the role of cyclic AMP.

Ca^{2+} ions have been proposed as the second messenger for a large number of external stimuli, some of which are listed in Table 6.1. In many cases the mechanisms involved in the alteration of Ca^{2+} levels are very poorly understood and the subject of much controversy. It is beyond the scope of this book to give a detailed consideration of all the systems where changes in cytoplasmic free Ca^{2+} levels have been suggested to act as second messenger for an external signal. Instead, the practical problems associated with determining changes in intracellular free Ca^{2+} levels will be considered. Then the system which is perhaps the best characterized, the response of liver cells to catecholamines, vasopressin, and angiotensin II, will be described. Finally, theoretical considerations arising from the known mechanisms for regulating the

Table 6.1. Agonists which change cytoplasmic free Ca^{2+}

Agonist	Tissue
α_1-Adrenergic	Liver, fat cell, muscle, nerve
Vasopressin	Liver
Angiotensin	Liver
5-Hydroxytryptamine	Insect salivary gland
Concanavalin A	Lymphocytes
Antibodies	Lymphocytes?
Growth factors	Various, depending on factor

cytoplasmic free Ca^{2+} concentration will be discussed. At this point it may be worth restating the criteria for the identification of a second messenger.

1 The messenger must affect the metabolic processes which are influenced by the hormone.
2 The hormone must change the cytoplasmic concentration of the second messenger in the appropriate direction.
3 A mechanism for removing the messenger must exist.
4 A causal relationship must be established between the binding of the hormone and the physiological response.

In the case of cyclic AMP, the enzymes which synthesize and remove cyclic AMP can be measured in broken cells, cyclic AMP levels can be measured easily, and the mechanism of action of cyclic AMP is well characterized. The mechanism by which Ca^{2+} affects the activity of proteins via the regulator protein, calmodulin, is also quite well characterized and will be described in Chapter 7. Cytoplasmic free Ca^{2+} levels, however, may be affected by alterations in the rate of either influx or efflux across the plasma membrane, mitochondrial membrane or endoplasmic reticulum membrane, or by the release of bound Ca^{2+} in the cytoplasm. Furthermore, since these are transport processes they can only be studied in intact membrane systems. Finally, it is difficult to measure the cytoplasmic free Ca^{2+} concentration accurately.

6B Methods for studying Ca^{2+} concentration changes in cells

6B1 Direct measurement of changes in cytoplasmic free Ca^{2+} concentration

The cytoplasmic free Ca^{2+} concentration is approximately 10^{-7} mol/l. This compares with a total cytoplasmic calcium of about $50-100$ μmol/l, a mitochondrial total Ca^{2+} of $2-3$ mmol/l, and an extracellular free Ca^{2+} of 1.3 mmol/l, the total Ca^{2+} in serum being 2.6 mmol/l. The cytoplasmic free Ca^{2+} concentration is therefore a very small proportion of the total calcium in the cell and, accordingly, very difficult to measure. Three types of method have been used to measure changes in the cytoplasmic free Ca^{2+} concentration.

Insertion of microelectrodes

It is possible to make electrodes which will respond specifically to the presence of a number of different ions, the obvious example being the

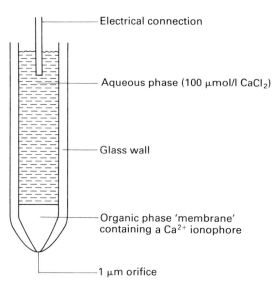

Fig. 6.1. A Ca^{2+}-sensitive electrode.

pH electrode which responds to H^+ ions. An ion-selective electrode depends upon a membrane which is selectively permeable to the ion under study, allowing a potential to be set up in response to the presence of the ion. The membrane may consist of a special glass, as in the case of a pH electrode, which will allow the passage of H^+ ions while excluding all others. In the case of a large ion such as Ca^+, a different strategy must be followed and the 'membrane' usually consists of a liquid, water-immiscible organic phase containing, in solution, a compound such as a chelating agent which will bind the ion selectively (Fig. 6.1). This will function as a selectively permeable membrane for the ion. Such electrodes can be made with diameters of as little as one micron at the tip, and they have been very useful for measuring cytoplasmic Ca^{2+} concentrations in large cells such as squid giant axons. Insertion of the electrode into smaller cells, such as a liver cell, has so far proved to be impractical, but techniques are improving all the time. Obviously there will always be uncertainty about the extent to which the cell is damaged by the insertion of the electrode. A second problem is the rapidity of the response of the electrode to changes in the ion concentration. Since the liquid phase 'membrane' is quite thick, response times can be several seconds. Changes in cytoplasmic Ca^{2+} over much shorter periods are likely to be important.

Opitical methods

The second main strategy has been the use of a wide variety of agents which either fluoresce when they bind Ca^{2+}, or change in optical absorbance. There are three main requirements for such a compound. It should have a binding constant for Ca^{2+} which is within the range of cytoplasmic free Ca^{2+} concentrations. Ideally, the binding should be half-saturated at a Ca^{2+} concentration between 10^{-7} and 10^{-6} mol/l. Furthermore, the agent must bind Ca^{2+} in preference to other ions, in particular Mg^{2+}. The Ca^{2+}-sensitive agent should enter the cell readily and be compartmentalized in the cytoplasm. The agent should not be toxic to the cell, and its entry into the cell should not disrupt the cell metabolism.

There are a number of agents which satisfy the first requirement. These include the proteins aequorin and obelin which are extracted from marine organisms, chemical dyes of which the most widely used is arsenoazo III and derivatives of the chelating agent EGTA such as Quin 2. The Ca^{2+}-sensitive proteins, aequorin and obelin, were the first to be used. These are highly specific for Ca^{2+} and once inside the cell appear to cause little metabolic disruption. However, being proteins they will not pass across the cell membrane and their successful use has been confined to large cells where they may be physically injected. Another approach is to fuse lipid vesicles containing the protein with the cells under study.

Chemical dyes, such as arsenoazo III, tend to be quite cytotoxic. They are also quite difficult to get into cells and tend to aggregate inside the cell which affects their optical absorbance. As a result, quantitative interpretation of the results is difficult. In general they have been more useful for the measurement of low levels of Ca^{2+} in extracellular media than for the measurement of free Ca^{2+} concentrations in the cytoplasm.

The most recent, and most promising, method is based upon derivatives of the Ca^{2+} chelator EGTA, of which Quin 2 has been the most widely used. EGTA is substituted with a quinoline group and a benzene ring (Fig. 6.2). As a result of this modification there is a large change in fluorescence on binding of Ca^{2+}. The substitution also lowers the Ca^{2+} dissociation constant from 10^{-11} mol/l for EGTA, to 10^{-7} mol/l for Quin 2. The Mg^{2+} dissociation constant is about 10^4-fold lower so that Mg^{2+} does not interfere with the signal to any extent. Thus from the fluores-

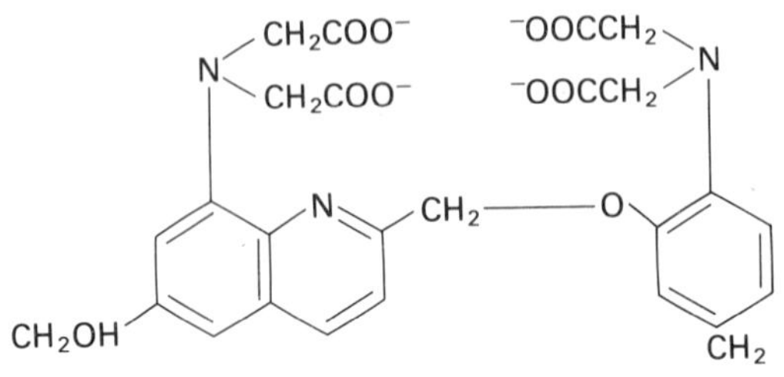

Fig. 6.2. Structure of Quin 2.

cence signal, the proportion of Quin 2 with Ca^{2+} bound can be measured, and from this, using the dissociation constant, the free Ca^{2+} concentration can be calculated.

An obvious problem with Quin 2 is the large negative charge of the four carboxyl groups. As might be expected, this prevents the molecule from passing across the plasma membrane. This problem was overcome by forming the acetomethoxy esters of the carboxyl groups. The esterified form is hydrophobic and passes readily across the plasma membrane. Once inside the cell, cytoplasmic esterase enzymes yield free Quin 2 and a mixture of acetate and formaldehyde. Free Quin 2, being highly charged, does not enter the mitochondria or endoplasmic reticulum. Because of the irreversible hydrolytic step, Quin 2 accumulates in the cytoplasm to a much higher concentration than the Quin 2 acetomethoxy ester concentration outside the cell. Quin 2 has been used successfully to measure the free concentration of Ca^{2+} in lymphocytes, platelets, and liver cells. Rapid changes in Ca^{2+} concentration can be seen in response to extracellular effectors such as the lectin, concanavalin A, working on lymphocytes, or α-adrenergic agonists working on liver cells. However, the approach is not without problems. Quin 2 is not cytotoxic but it has been found in some cell types to affect the metabolism of the cell. In lymphocytes, the rate of glycolysis is increased and Quin 2 itself is a weak mitogenic agent. These difficulties are probably not insuperable. To get a good optical signal, a cytoplasmic concentration of about 1 mmol/l Quin 2 is required. This is partly because the fluorescence yield is not particularly high, and partly because the wavelengths for maximum excitation and emission coincide with the background absorbance and fluorescence of the cell. It should be possible to synthesize new molecules along the same lines avoiding these problems. In this case, the concentration required in the cytoplasm would be much lower with much less likelihood of metabolic disruption.

Null point titration

The idea behind null point titration is outlined in Fig. 6.3. The cells under study are suspended in media containing varying concentrations of Ca^{2+}. The incubations are otherwise exactly the same. A range of Ca^{2+} concentrations is chosen to include the expected cytoplasmic free Ca^{2+} concentration, and the extracellular concentration is measured by a method which gives a rapid response. The dye arsenoazo III is often used or, alternatively, a Ca^{2+}-sensitive electrode could be used. At a chosen time, the plasma membrane is made freely permeable to Ca^{2+} without disrupting the mitochondrial membrane or endoplasmic

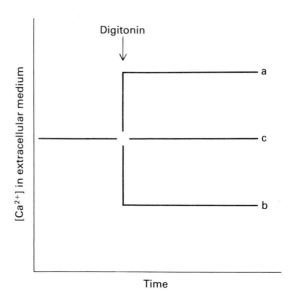

Fig. 6.3. Null point titration. The Ca^{2+} concentration in the external medium is measured with a dye or a Ca^{2+} electrode. The plasma membrane is disrupted with digitonin which leaves the intracellular membranes intact. If the intracellular concentration is greater than the extracellular concentration, Ca^{2+} release causes an increase in the measured Ca^{2+} concentration (a). If the intracellular concentration is less than the extracellular concentration, dilution causes a decrease in the measured Ca^{2+} concentration (b). If the intracellular and extracellular concentrations are the same, there will be no change in the measured Ca^{2+} concentration (c).

reticulum membrane. This is usually done with the detergent digitonin, which interacts preferentially with cholesterol, which is only found in the plasma membrane and not in the other membranes of the cell. If the concentration of Ca^{2+} outside the cell is lower than the cytoplasmic free Ca^{2+} concentration, it will increase when the plasma membrane becomes freely permeable to Ca^{2+}. If, on the other hand, the Ca^{2+} concentration outside the cell is higher than the cytoplasmic free Ca^{2+} concentration, then the Ca^{2+} concentration measured will decrease when the plasma membrane is made permeable to Ca^{2+}. If both concentrations are the same there will be no change in Ca^{2+} concentration when the plasma membrane is disrupted, and this can be used as a measure of the cytoplasmic free Ca^{2+} concentration.

The method has the advantage that it is non-invasive, in that no foreign material is introduced into the cell before the measurement. On the other hand, the cells must be exposed to external Ca^{2+} concentrations three to four orders of magnitude below the normal extracellular concentration of 1.3 mmol/l, and this may well affect the cellular response. Furthermore, the cytoplasmic Ca^{2+} concentration can only be measured at a single time point in each incubation. Optical methods or electrodes give a continuous measurement.

None of the methods for measurement of cytoplasmic free Ca^{2+} concentration are entirely satisfactory. Perhaps the best strategy is to use more than one method. Where this has been done the results are usually in quite good agreement, most cell types appearing to have a resting Ca^{2+} concentration between 1 and 2×10^{-7} mol/l.

6B2 Ca^{2+} flux studies

The second general type of approach to the problem of determining changes in intracellular free Ca^{2+} concentrations is to measure the flux of Ca^{2+} into or out of the cell. Attempts are then made to extrapolate from changes in the flux rate to changes in the intracellular Ca^{2+} levels. Obviously this is a less direct method than the measurement of cytoplasmic free Ca^{2+} concentrations. There are two basic approaches. Both may be applied to isolated cells, to pieces of intact tissue, or to perfused whole organs. Most methods for direct measurement of cytoplasmic Ca^{2+} require that the cells be isolated from the tissue.

^{45}Ca flux measurements

Either ^{45}Ca influx into the cell or ^{45}Ca efflux from a cell which has been preloaded with ^{45}Ca can be measured. In many cell types, hormones have been shown to affect both influx and efflux rates. The method has the advantage that, in contrast to most other methods for examining Ca^{2+} changes in cells, it does not perturb the normal cellular environment; physiological levels of extracellular free Ca^{2+} can be used, and the cell is not exposed to any mechanical or chemical invasion.

A change in ^{45}Ca flux in response to an effector is an excellent indication that the balance of Ca^{2+} concentrations inside the cell is changing. However, it is very difficult to determine the mechanism producing the change or, indeed, even whether the free cytoplasmic Ca^{2+} concentration is increasing or decreasing. This is largely because the total cell Ca^{2+} does not equilibrate with ^{45}Ca within the time that it is possible to maintain the viability of most tissues. Thus the specific radioactivity of the many different pools of Ca^{2+} in the cell is not the same. When the release of ^{45}Ca from preloaded cells is measured, it is assumed that ^{45}Ca is released into the medium from the cytoplasm across the plasma membrane, and that the rate of release is affected by the cytoplasmic free Ca^{2+} concentration. If the effector causes the

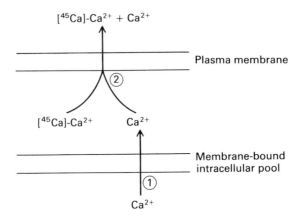

Fig. 6.4. Effect of isotope dilution from an intracellular source on ^{45}Ca efflux. This represents an extreme case in which the intracellular Ca^{2+} pool is not labelled with ^{45}Ca at all. If, in the unstimulated state, efflux from the cell (2) is made up of equal amounts of [^{45}Ca]-Ca^{2+} and unlabelled Ca^{2+} from the intracellular pool, a threefold increase in release from the pool (1) together with a twofold increase in release from the cell (2) would give a 33% reduction in ^{45}Ca release, in spite of a doubling of total Ca^{2+} efflux.

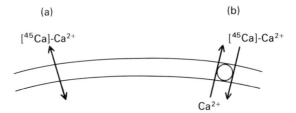

Fig. 6.5. Effect of Ca^{2+}/Ca^{2+} exchange on [^{45}Ca]-Ca^{2+} influx. (a) Net influx or efflux. This affects both intracellular Ca^{2+} content and radioactive labelling. (b) Ca^{2+}/Ca^{2+} exchange. This affects only labelling with ^{45}Ca.

mobilization of substantial amounts of Ca^{2+} from an intracellular source, the cytoplasmic Ca^{2+} concentration will go up. However, if the Ca^{2+} entering the cytoplasm from the intracellular source is of a much lower specific radioactivity than the Ca^{2+} already in the cytoplasm, the dilution effect may be sufficient to reduce the rate of release of ^{45}Ca from the cell even though the net Ca^{2+} flux from the cell may be increased (Fig. 6.4).

Similar problems can arise with ^{45}Ca influx measurements. Most plasma membranes contain a Na$^+$/Ca^{2+} exchanger (see Section 5C2) which also has the ability to exchange Ca^{2+} for Ca^{2+}. If the cell is exposed to ^{45}Ca outside, the rate of influx of [^{45}Ca]-Ca^{2+} can be increased, either by increasing the rate of net influx of Ca^{2+} or by increasing the rate of Ca^{2+}/Ca^{2+} exchange. If the rate of exchange is increased, the rate of ^{45}Ca influx can increase without any increase in net influx of Ca^{2+}. Indeed there can be net efflux of Ca^{2+} from the cell and, provided that Ca^{2+}/Ca^{2+} exchange is increased enough, there will be a net influx of [^{45}Ca]-Ca^{2+} (Fig. 6.5). The interpretation of ^{45}Ca flux data, therefore, requires a great deal of caution.

Direct measurement of Ca^{2+} release

As an alternative to the measurement of ^{45}Ca fluxes, Ca^{2+} release can be measured directly. This simply involves the measurement of the appearance of Ca^{2+} in the medium and any sufficiently sensitive method can be used. It avoids the problem of the specific activity of ^{45}Ca, but the amount of Ca^{2+} released will usually be far too small to be detected against the background of a normal physiological extracellular Ca^{2+} concentration. As a result, this type of experiment usually employs a low Ca^{2+} medium rather than a physiological medium, and the interpretation of any changes observed is still difficult.

6C The effects of hormones on liver cells

The production of glucose during starvation to maintain blood glucose levels is one of the major functions of the liver. There are two sources of glucose: the breakdown of stored glycogen and synthesis *de novo*, or gluconeogenesis from lactate, pyruvate and amino acids. Both processes are activated by glucagon, vasopressin, angiotensin II, and catecholamines. The glucagon effect is mediated by a rise in cyclic AMP acting through the cyclic AMP-dependent protein kinase, as described in Chapters 3 and 4. The effects of vasopressin and angiotensin II appear to be independent of changes in cyclic AMP. The effects of catecholamines were thought to be mediated in the cyclic AMP pathway acting via β receptors which are present in liver. It is now thought that in many mammals the effects of catecholamines are mediated via a separate class of receptors—the α_1 receptors.

6C1 Catecholamine receptors

The occurrence of different types of catecholamine receptors has been mentioned a number of times in previous chapters. At this point it is appropriate to consider the question in more detail.

Pharmacological criteria

Catecholamines are relatively simple molecules and easy to modify chemically, and the synthesis of analogous molecules is also relatively easy. There are also good clinical reasons to either stimulate catecholamine responses, for example in the treatment of asthma, or to inhibit catecholamine responses, as in the treatment of hypertension. These two factors have led to the development of a very large number of different

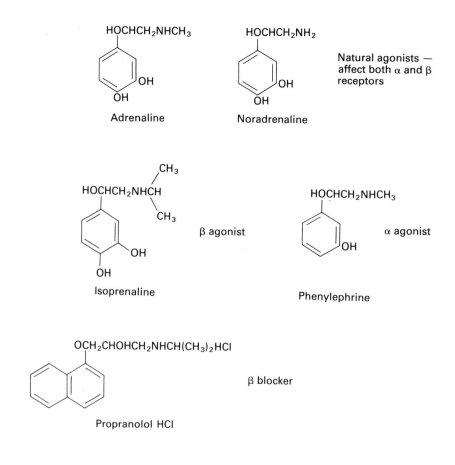

Fig. 6.6. Structures of some commonly used adrenergic agonists and blockers.

analogues of adrenaline and noradrenaline. The structure of a few of these is shown in Fig. 6.6. Some of these analogues mimic the effects of adrenaline, while some block the response. Different physiological responses were classified according to the ability of different analogues to either produce or block the response, and eventually this led to identification of four classes of receptor: α_1, α_2, β_1, and β_2. The specificity of adrenergic agonists and blockers is shown in Table 6.2.

Table 6.2. Specificity of artificial adrenergic agonists and blockers

Isoprenaline	β agonist
Phenylephrine	α_1 and α_2 agonist
Clonidine	α_2 and α_1 agonist
Methoxamine	α_1 agonist
Propranolol	β blocker
Dihydroalprenolol	β blocker
Phenoxybenzamine	α_1 and α_2 blocker
Yohimbine	α_2 blocker
Prazosin	α_1 blocker

Biochemical criteria

Later it was found that the different pharmacological classes of receptor defined by chemical specificity also had different mechanisms of action. β receptors occur extensively in both nerve endings and peripheral tissues. Both β_1 and β_2 receptors function by the activation of adenylate cyclase. α receptors were originally described in the nervous system. α_1 receptors were found to occur on post-synaptic membranes while α_2 receptors were found on both pre-synaptic and post-synaptic membranes. Both occur in peripheral tissues as well. α_2 receptors act by the direct inhibition of adenylate cyclase in the post-synaptic membrane and in the plasma membrane of peripheral tissue cells. Pre-synaptic α_2 receptors are responsible for feedback inhibition of noradrenaline release, but the mechanism has not been characterized. α_1 receptors act by causing an increase in the cytoplasmic free Ca^{2+} concentration, and this is the mechanism which concerns us in this chapter.

Many hormones have more than one receptor type, each having a different mechanism of action. Vasopressin, angiotensin II and dopamine all have at least two classes of receptor: one which activates

94 *Chapter 6*

adenylate cyclase and one which modulates cytoplasmic free Ca^{2+} levels. Prostaglandins and adenosine may either activate or inhibit adenylate cyclase. In the case of catecholamines the receptor heterogeneity is well characterized, but it seems likely to be a fairly general phenomenon. Insulin appears to be an exception to this, in that a single type of receptor appears to be linked to a number of different mechanisms, rather than a number of different receptor types, each being responsible for a different mechanism.

6D Ca^{2+} as a second messenger in liver: evidence for the involvement of Ca^{2+} rather than cyclic AMP

Glucagon clearly exerts its effects on liver cells through activation of adenylate cyclase and a rise in cyclic AMP. There is no evidence that vasopressin or angiotensin affect levels of cyclic AMP in the liver, so an alternative mechanism must be sought to explain the increase in glucose release in response to these hormones. Adrenaline does activate adenylate cyclase in rat liver, the usual source for liver cell preparations, and

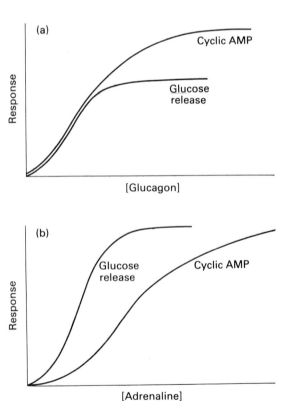

Fig. 6.7. Stimulation of glucose release and increase in cyclic AMP levels in liver cells in response to glucagon and adrenaline. (a) With glucagon, the stimulation of glucose release correlates well with the initial rise in cyclic AMP. (b) With adrenaline, glucose release is stimulated at hormone concentrations which have little effect on cyclic AMP levels.

does raise cyclic AMP levels. For a long time it was thought that this was responsible for the effect of adrenaline in stimulating glucose release. However, adrenaline was a much less effective activator of adenylate cyclase in rat liver than glucagon, while being equally effective in activating glucose release. In the early 1970s it was found that α-adrenergic analogues of adrenaline also stimulate glucose output from liver. On examination of the effect of α- and β-agonists on glucose output, it was found that the α-agonists were more effective and that the correlation of rate of glucose release with cyclic AMP concentration was very poor (Fig. 6.7). Thus it became apparent that for adrenaline, as well as for vasopressin and angiotensin II, a cyclic AMP-independent mechanism must exist. Glucagon is physiologically the most important hormone controlling liver glucose metabolism. However, the effects of vasopressin and α_1-adrenergic agonists provide a useful model system in which to study the mechanism of action of Ca^{2+}-mobilizing hormones.

6D1 Dependence on extracellular Ca^{2+}

The first evidence for a role for Ca^{2+} in the action of vasopressin, angiotensin and catecholamines, came from studies on the effects of removing extracellular Ca^{2+}. If liver cells were suspended in a Ca^{2+}-free medium, containing the chelating agent EGTA, for about one hour, the response to adrenaline, vasopressin or angiotensin II was lost. The hormone response was restored by the restoration of Ca^{2+} to the medium, and this also required about one hour. If the hormones were added shortly after the removal of Ca^{2+} from the extracellular medium, the liver cells responded more or less normally to the addition of any of these hormones with increased glucose release (Fig. 6.8). There are two possible explanations for these observations. The first assumes that the free Ca^{2+} concentration in the cytoplasm changes in response to the hormone, and that this is an essential part of the mechanism of action of the hormone. If this is the case, it seems that the increase in cytoplasmic free Ca^{2+} is derived from an intracellular source since the hormones will work in the absence of extracellular Ca^{2+}. If, however, the intracellular calcium is depleted by prolonged exposure to calcium-free medium, the hormone effect is lost due to the depletion of the intracellular source for the rise in Ca^{2+}. The second interpretation of the data starts from the assumption that Ca^{2+} is not the cytoplasmic signal for the action of the hormone. The lack of any effect of short-term exposure to Ca^{2+}-free medium is consistent with this. The loss of the hormone response, or prolonged exposure to Ca^{2+} free-medium, can then be explained by suggesting that extensive Ca^{2+} depletion of the cell causes a general

96 Chapter 6

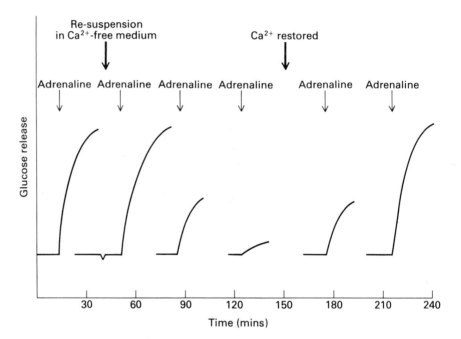

Fig. 6.8 Effect of Ca^{2+}-free medium on the stimulation of glucose release from liver cells by α-adrenergic agonists. This shows the effect of adding adrenaline to liver cells after exposure to Ca^{2+}-free medium for varying periods of time. The response is lost after about one hour, and returns after about one hour of re-exposure to Ca^{2+}.

impairment of cell function. The results of this approach are therefore inconclusive and more convincing evidence of an involvement of Ca^{2+} in the mechanism of action of the hormone is needed.

6D2 Measurement of cytoplasmic free Ca^{2+}

Null point titration experiments and experiments with the fluorescent Ca^{2+} indicator, Quin 2, show that the cytoplasmic free Ca^{2+} concentration in liver cells increases in response to the three hormones. The hormones also cause large changes in ^{45}Ca fluxes, and in the net release of calcium from isolated liver cells and from intact perfused liver. The problems associated with this type of study have been discussed above (see Section 6C), and it is true that individually the results of all these approaches have to be treated with caution. Taking all the data together, however, the evidence that the three hormones cause a rise in cytoplasmic free Ca^{2+} levels in the liver is very strong. The source of the Ca^{2+}, and the mechanism of the increase, is much more controversial.

6E Effects of Ca^{2+} on liver metabolism

The next question to consider is whether there is any evidence that a rise in cytoplasmic free Ca^{2+} will result in the activation of glucose release. There are two approaches to this question: one is to raise the concentration of Ca^{2+} in the cytoplasm by an artificial stimulus and examine the effect of this on glucose output. This has been achieved using the Ca^{2+} ionophore A23187 (Fig. 6.9), a lipid-soluble, Ca^{2+}-chelating agent which has the effect of allowing Ca^{2+} to pass freely across the plasma membrane and also across the other membranes of the cell. If Ca^{2+} acts as second messenger for hormones which promote glucose release, then in the presence of the ionophore, glucose release should be sensitive to the external Ca^{2+} concentration. It was found that increasing Ca^{2+} outside liver cells from 10^{-7} to 10^{-6} mol/l in the presence of A23187 caused a marked increase in glucose release. High concentrations of Ca^{2+} in the presence of A23187 were toxic to the cell, probably due to the mitochondria becoming overloaded with Ca^{2+}.

The second approach is to look for direct effects of Ca^{2+} on the enzymes involved in glucose release. In the case of glycogen breakdown, there is a very clear-cut activation of glycogen phosphorylase kinase by Ca^{2+}, leading to increased phosphorylation and activation of phosphorylase, the enzyme responsible for glycogen breakdown (see Chapters 14 and 15). There is also evidence that Ca^{2+} activates gluconeogenesis, but the mechanism is less well characterized.

Two of the criteria for establishing the role of Ca^{2+} as a second messenger are well supported by the experimental evidence. There is good evidence that hormones change cytoplasmic free Ca^{2+} levels and good evidence that an increased cytoplasmic free Ca^{2+} will activate glucose release. The third criterion is the existence of a mechanism for the removal of the messenger, and the presence of a number of carriers able to transport Ca^{2+} out of the cytoplasm is well established. However,

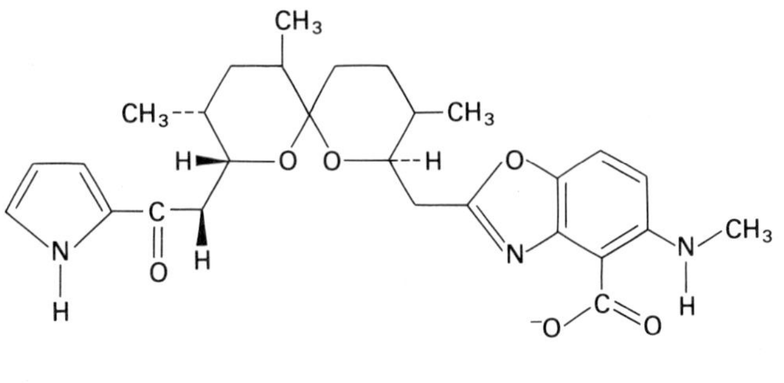

Fig. 6.9 Structure of A23187.

we do not have a detailed description of the whole process by which hormones make use of Ca^{2+} to produce a cellular response. In particular, the source of the rise in cytoplasmic Ca^{2+} is controversial.

6F Source of the rise in cytoplasmic free Ca^{2+}

There are three possible sources for the rise in cytoplasmic free Ca^{2+}.
1   The extracellular space.
2   An intracellular organelle such as the mitochondrion or microsome.
3   A bound calcium pool in contact with the cytoplasmic water.

Cytoplasmic free Ca^{2+} levels might be increased by activating influx or inhibiting efflux across the plasma membrane, or by activating efflux or inhibiting uptake in an organelle. Finally, Ca^{2+} may be released into the cytoplasm from binding sites, for example on membrane surfaces. From our knowledge of the mechanism available for controlling cytoplasmic free Ca^{2+} levels, we can define certain constraints.

6F1 Time course of the Ca^{2+} response

The first question to be asked is 'How long does the Ca^{2+} level need to be elevated to produce the physiological response?' The rise in Ca^{2+} may need to persist for as long as the hormone is present, and for as long as the response takes place. Alternatively, the rise in Ca^{2+} may be transitory and the physiological response may persist after the Ca^{2+} concentration returns to the ground state. These two possibilities each raise different theoretical considerations.

Persistent Ca^{2+} rise

If the increase in cytoplasmic free Ca^{2+} persists for as long as the hormone is present, then the plasma membrane must play a key role in the response. Intracellular calcium stores are finite, and in the absence of any modulation of Ca^{2+} influx and efflux across the plasma membrane, an increase in cytoplasmic free Ca^{2+} cannot be maintained indefinitely. On the other hand, there are constraints upon the possibility of net influx or Ca^{2+} across the plasma membrane. If this were to lead to an increase in the cytoplasmic free Ca^{2+} concentration to a level above the mitochondrial buffering point, then persistent accumulation of Ca^{2+} by the mitochondria would lead to the uncoupling of ATP synthesis. Ultimately, the mitochondria would become overloaded with calcium and stop functioning. Thus, for a prolonged rise in the cytoplasmic free Ca^{2+} concentration, two conditions must be satisfied.

First, there must be some modulation of the Ca^{2+} fluxes across the plasma membrane, and second, the mitochondrial buffering point must not be exceeded. This does not preclude the possibility that the mitochondrial buffering point might change in response to the hormone. The actual source of the Ca^{2+} may also be intracellular. An inhibition of Ca^{2+} efflux across the plasma membrane would maintain an increase in the cytoplasmic free Ca^{2+}, whatever the original source of net influx of calcium into the cytoplasm.

Transitory Ca^{2+} rise

In the case of a transitory rise in cytoplasmic Ca^{2+} in response to a hormone, there is no constraint upon possible sources of Ca^{2+}. Any of the possible pools of calcium—the extracellular space, bound cytoplasmic calcium, mitochondrial or endoplasmic reticulum calcium—could allow a short rapid flux of Ca^{2+} into the cytoplasm. The cytoplasmic free Ca^{2+} would then be restored to the ground state by the pumping of Ca^{2+} out of the cell across the plasma membrane. In many cases, however, the metabolic response persists for as long as the hormone is present. In this case we must explain how a transient rise in the second messenger, Ca^{2+}, causes a persistent response to the hormone. There are a number of possibilities to consider.

1 The activation of the metabolic process is mediated by Ca^{2+}. If, at the same time, the hormone inhibits the deactivation process, then a transient rise in Ca^{2+} will complete the activation and the system will remain active as long as the reversal process is inhibited (Fig. 6.10). This implies the existence of another messenger to initiate the inactivation process.

2 Ca^{2+} acts through the binding protein calmodulin. This will be discussed in detail in Chapter 7. There is evidence that the interaction of calmodulin with target enzymes causes an increase in the binding affinity for Ca^{2+}. Thus, as the free Ca^{2+} falls, the calmodulin, as long as it is associated with its target protein, will retain its bound calcium and the target enzyme will still be activated. Reversal of the activation will then require a reduction in the level of cytoplasmic free Ca^{2+} below the original resting level, or some other active intervention to affect the affinity of calmodulin binding to protein.

There is good evidence that the rise in cytoplasmic Ca^{2+} in response to the hormone is transitory, and that the source is intracellular. The Ca^{2+} indicator, Quin 2, has been used in liver cells to measure rapid changes in cytoplasmic free Ca^{2+} concentrations in response to hormones. Vasopressin and α_1-adrenergic agonists caused a rise in Ca^{2+}

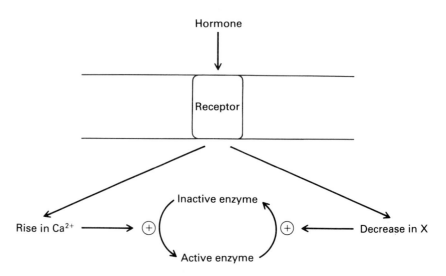

Fig. 6.10 Model of independent mechanisms for stimulation and inhibition of response to a hormone. As a consequence of binding to its receptor, the hormone causes an increase in Ca^{2+} which promotes the conversion of the enzyme to its active form. This would be potentiated by a hormone-induced decrease in a signal (X) which promotes the conversion of active enzyme to inactive enzyme. This would also have the effect of enabling a persistent activation in response to a transient Ca^{2+} rise if the effect of the hormone on X persists after the effect on Ca^{2+}.

which was detectable after two seconds and maximal after six seconds. The Ca^{2+} levels then declined over a period of about five minutes. If the cells were first exposed to an α_1-adrenergic agonist, subsequent addition of vasopressin caused no further increase in Ca^{2+} as long as the α_1-adrenergic agonist was still present. However, if the effect of the α_1-agonist was removed by adding a suitable blocker, vasopressin was able to cause a second increase in cytoplasmic Ca^{2+}, but only after a lag of about fifteen minutes. This suggests that the pool, which is the source of the Ca^{2+} rise, was depleted by the first hormone which also prevented the pool being restored, indicating that the hormone does have effects on the cell which persist after the rise in Ca^{2+}. After removal of the first hormone, the pool would refill, a process taking about fifteen minutes. It is clear that the Ca^{2+} pool must be intracellular since it is impossible to deplete the extracellular pool.

Direct flux studies have been performed in which net release or uptake of Ca^{2+} is measured, using both isolated liver cells and perfused liver. Most groups report a rapid net efflux of Ca^{2+} within a minute of addition of adrenaline, reaching a maximum after 5–8 minutes. By 11 minutes, Ca²⁺ efflux returns to basal levels and is then followed by a slow re-uptake phase. The activity of the enzyme phosphorylase kinase, which is sensitive to Ca^{2+}, can be used as an indicator of changes in cytoplasmic free Ca^{2+} (see Section 14F). Activation of phosphorylase kinase follows the time course of the initial increase in efflux, not the later influx phase (Fig. 6.11). This suggests again that the source of the rise in cytoplasmic Ca^{2+} is intracellular, leading to increased Ca^{2+} efflux. The uptake phase is taken to represent replenishment of cell

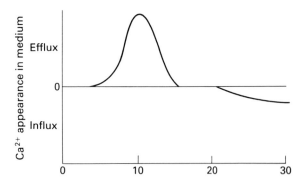

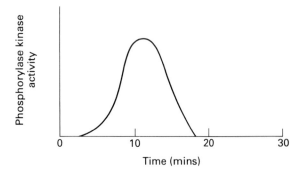

Fig. 6.11 Relationship of phosphorylase kinase activity to Ca^{2+} efflux and influx from liver cells. Activation of phosphorylase kinase corresponds to the early efflux phase rather than to the later influx phase.

Ca^{2+}. In similar experiments, glucose release has also been measured. A rise can be detected within 90 seconds and is maximal within 3–4 minutes. However, the stimulation of glucose release persists long after the stimulation of Ca^{2+} efflux has stopped (Fig. 6.12).

To summarize the conclusions from these experiments: it appears that the hormone causes a transitory rise in cytoplasmic Ca^{2+} from an intracellular source. The physiological response, however, persists after the Ca^{2+} rise, and the hormone appears to have other effects on Ca^{2+} fluxes in the cell after the initial mobilization.

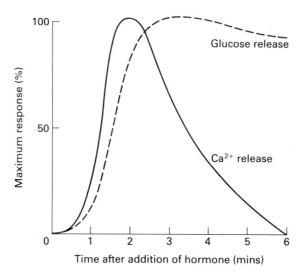

Fig. 6.12 Time course of Ca^{2+} and glucose release from rat liver.

6G The nature of the intracellular Ca^{2+} pool

There are three likely candidates for an intracellular source for the rise in cytoplasmic free Ca^{2+} in response to hormones: the mitochondria, the endoplasmic reticulum, and bound Ca^{2+} in the cytoplasm. It has proved to be quite difficult to establish which pool of Ca^{2+} is the source.

As discussed above (see Section 6C1), ^{45}Ca fluxes are difficult to interpret. However, the usefulness of the technique can be increased by computer analysis of rates of efflux from preloaded cells. The most extensive study used a computer model which assumed three pools: the extracellular space, a small, rapidly exchangeable pool taken to represent cytoplasmic Ca^{2+}, both free and bound, and a slowly exchangeable pool taken to consist of the mitochondria and microsomes together. Hormones increase the size of the pool taken to represent the cytoplasm, and decrease the size of the membrane-bound pool, leading to the conclusion that the major source of the rise in cytoplasmic free Ca^{2+} is an intracellular membrane-bound pool such as the mitochondria or endoplasmic reticulum. This leads to the conclusion that Ca^{2+} is not the second messenger for the hormone. Since the hormone does not enter the cell, another signal molecule must be produced in response to hormone binding which crosses the cytoplasm to stimulate Ca^{2+} release from the intracellular organelle.

Early attempts to establish which organelle is responsible for the Ca^{2+} release favoured the mitochondria. Methods were developed for the rapid isolation of mitochondria from liver cells, and there are a number of reports that treatment of the intact cell with α-adrenergic agonists results in a reduction of mitochondrial total calcium of between thirty and seventy per cent within one or two minutes. At longer time periods of around thirty minutes α-adrenergic agonists cause a rise in total mitochondrial calcium. These results are consistent with the conclusions of ^{45}Ca flux studies and suggest that the mitochondria are the source of the rise in cytoplasmic free Ca^{2+}. This would require a second messenger which would either increase the rate of efflux from the mitochondria or decrease the rate of influx.

However, in the case of the mitochondria, it seems likely that in the resting state calcium efflux is limited by the availability of free Ca^{2+} inside the mitochondria. This follows from the fact that the mitochondrial 'buffering' point appears to be at a higher free Ca^{2+} concentration than the resting free Ca^{2+}. So Ca^{2+} release from the mitochondria would require a substantial change in the rate of either the efflux carrier or the influx carrier. There is some evidence that this may happen. Mitochondria isolated from heart, previously exposed to α-adrenergic agonists, display increased activity of the uptake carrier. If liver is treated with glucagon β-adrenergic agonists or cyclic AMP before isolating the mitochondria, the activity of a Na^+-dependent efflux carrier is increased so that it accounts for about fifty per cent of the total efflux rate. In control mitochondria, the activity of the Na^+-dependent carrier is insignificant in comparison to Na^+-independent efflux. The physiological significance of these effects is not known. It seems likely that the effect of cyclic AMP, and of hormones which raise cyclic AMP in liver, may be concerned with a modulation of the Ca^{2+} signal by cyclic AMP. The activation of the uptake carrier in heart is perhaps more likely to be involved in the removal of the Ca^{2+} signal rather than in its generation. If the activity of the mitochondrial carriers does not change, then bound calcium inside the mitochondria would have to be mobilized. This would require a rather complex mechanism whereby the hormone binding to the receptor generates a second messenger which interacts with the mitochondria in such a way as to mobilize bound intramitochondrial calcium.

Recently, attention has focused on the endoplasmic reticulum as a source of the rise in cytoplasmic Ca^{2+}. By making the plasma membrane freely permeable to Ca^{2+} with digitonin (see Section 6B1) it is possible to analyse the intracellular pools in more detail. This approach revealed the presence of a small membrane-bound pool with an affinity for Ca^{2+} of about 10^{-7} mol/l which was thought to be the endoplasmic reticulum, and a large pool with an affinity of 10^{-6} mol/l for Ca^{2+} which was thought to be the mitochondria. The small pool was depleted by pre-incubation with Ca^{2+}-mobilizing hormones, suggesting that the endoplasmic reticulum may be the source of the rise in cytoplasmic free Ca^{2+}. In the last year or so, very good evidence in support of this view has accumulated in that a second messenger, capable of promoting release of Ca^{2+} from the endoplasmic reticulum, has been identified. The messenger is inositol trisphosphate which is produced as a result of breakdown of phosphatidylinositol 4,5-bisphosphate in the plasma membrane. This is activated by Ca^{2+}-mobilizing hormones and is described in detail in Chapter 9.

Many of the details of this process remain to be characterized. Most importantly, the mechanism by which hormones activate hydrolysis of phosphatidylinositol 4,5-bisphosphate to release inositol trisphosphate, and by which inositol trisphosphate causes the release of Ca^{2+} from the endoplasmic reticulum, are not understood. However, it is likely that inositol trisphosphate mediates the initial rise in Ca^{2+} in response to Ca^{2+}-mobilizing hormones.

Suggestions for further reading

Akerman K. & Nicholls D.G. (1983) Physiological and biochemical aspects of mitochondrial calcium transport. *Rev Physiol Biochem Pharmacol*, **95**, 152−159.

Exton J.H. (1980) Mechanisms involved in α-adrenergic phenomena. *Am J Physiol*, **238**, E3−E12.

Rasmussen H. & Waismann D.M. (1983) Modulation of cell function in the calcium messenger system. *Rev Physiol Biochem Pharmacol*, **95**, 111−151.

Thomas M.V. (1982) *Techniques in Calcium Research*. Academic Press.

Williamson J.R., Cooper R.H. & Hoek J.B. (1981) Role of calcium in the hormonal regulation of liver metabolism. *Biochim Biophys Acta*, **639**, 243−295.

Chapter 7 Mechanisms of action of calcium as a regulator

7A Introduction

Ca^{2+} ions regulate the activity of a large number of different enzymes in the cytoplasm and also a few intramitochondrial enzymes. The mechanisms of action appear to be different. In the cytoplasm, Ca^{2+} first binds to one of a number of calcium-binding proteins and it is the calcium–protein complex which modulates enzyme activity. In the mitochondria, Ca^{2+} ions appear to affect enzyme activity by a direct binding interaction.

7B Calcium-dependent regulator proteins

Many processes in the cell are affected by Ca^{2+} ions. In some cases, for example muscle contraction or glycogen breakdown, the enzymes involved were found to respond to Ca^{2+} concentrations close to those expected to occur in the cytoplasm. In other cases, such as lipolysis or

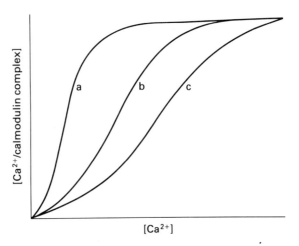

Fig. 7.1 Effect of dilution of cell contents on Ca^{2+}/calmodulin complex formation. Dilution of extract increases from a to c. The binding reaction can be simply described by the equation:

calmodulin + Ca^{2+} $\rightleftharpoons$ calmodulin/Ca^{2+} complex.

The concentration of the complex is governed by the binding constant:

$$\frac{[\text{calmodulin}/Ca^{2+} \text{ complex}]}{[\text{calmodulin}] [Ca^{2+}]}$$

Thus as the concentration of calmodulin is reduced by dilution of the cell contents, the concentration of Ca^{2+} needed to achieve a given concentration of Ca^{2+}/calmodulin complex will increase. Dilution will also affect the interaction of the Ca^{2+}/calmodulin complex with the responding enzyme tending to increase the Ca^{2+} requirement further.

cyclic AMP breakdown in brain, the enzyme concerned could be shown to respond to Ca^{2+}, but at much higher concentrations. The reason for this discrepancy became apparent with the discovery of Ca^{2+}-binding proteins. Muscle fibres contain a tightly bound protein, troponin C, which mediates the effects of Ca^{2+} on muscle. The effects of Ca^{2+} on many cytoplasmic enzymes are mediated by the soluble Ca^{2+}-binding protein calmodulin. Thus the activity of the enzyme depends upon the concentration of the calcium/calmodulin complex and not directly upon the free Ca^{2+} concentration. On disruption of a cell, the cytoplasmic contents are diluted reducing the concentration of both calmodulin and the target enzyme. As a result, a higher concentration of Ca^{2+} will be needed to achieve a particular concentration of the Ca^{2+}/calmodulin complex (see Fig. 7.1). This explains why many calcium-sensitive enzymes require high concentrations of Ca^{2+} in broken cell extracts.

A number of Ca^{2+}-binding proteins have been discovered. Troponin C is well characterized as the binding protein mediating the action of calcium in heart and skeletal muscle. Other muscle proteins are leiotonin, which has been suggested to serve a similar function in smooth muscle, and parvalbumen, which is found in skeletal muscle. The function of leiotonin and parvalbumen is poorly understood. Calmodulin is a cytoplasmic protein which is ubiquitous in vertebrate cells and has been reported in numerous non-vertebrate species including some protozoa. It is responsible for mediating the action of Ca^{2+} on many enzymes and is therefore central to the mechanism of action of calcium-mobilizing hormones.

7C Calmodulin

7C1 Structure and physical properties

Calmodulin has a molecular weight of 16 680. Approximately one-third of the amino acids are either glutamate or aspartate, and the isoelectric point is accordingly very low at pH 4.2. Another unusual feature is the complete lack of tryptophan or cysteine residues in the sequence. The amino acid sequences are known for four vertebrate and one invertebrate calmodulin. The sequence is very highly conserved, as might be expected for a protein with a major regulatory role. There are only seven amino acid changes out of a total of 148 when calmodulin, from the most primitive source which has been sequenced, is compared with the protein from bovine brain. The changes which do occur are near the N-terminal or C-terminal ends of the polypeptide chain, or are conser-

Mechanisms of action of calcium as a regulator

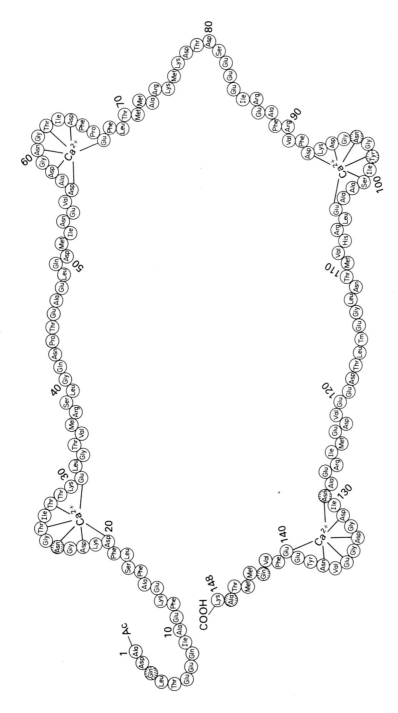

Fig. 7.2. Amino acid sequence of calmodulin showing Ca^{2+}-binding domains. (Reproduced, with permission, from Vanaman T.C. (1980) Calmodulin. In *Calcium and Cell Function* (Cheung W.Y., ed.), vol. 1, chap. 3, Academic Press. New York.)

vative. The terminal sequences of the protein are not thought to be functionally important.

Calmodulin contains four Ca^{2+}-binding domains, each consisting of two regions of α-helix separated by a Ca^{2+}-binding loop (Fig. 7.2). The four domains are very similar to each other suggesting that they originally arose through gene duplication, but no two domains are completely identical. The three linking peptides and the amino terminal peptide show large differences in sequence, but the very high degree of conservation of the sequence from one species to another suggests that all parts of the molecule, and not just the calcium-binding domains, are functionally important. Hydrodynamic and spectroscopic studies indicate that calmodulin is a compact globular protein with a high content of secondary structure—about 30% alpha-helix and 20% beta-pleated sheet.

7C2 Ca^{2+} binding

Calmodulin contains four high-affinity binding sites for Ca^{2+} with dissociation constants between 10^{-6} and 10^{-5} mol/l. This supports the conclusion based on the amino acid sequence that there are four Ca^{2+}-binding domains. There appear to be two classes of site of different affinity or, alternatively, negative cooperativity between the sites. The precise values of the Ca^{2+}-binding constants have still not been established, but it is clear that both monovalent and divalent cations can affect calcium binding. Mn^{2+} acts as an analogue of Ca^{2+} with about ten per cent of the affinity. Mg^{2+} competes with Ca^{2+} for binding but, in contrast to Mn^{2+}, does not produce any conformational change in the protein. Obviously we would like to know the extent of Ca^{2+} binding to calmodulin under the conditions prevailing in the intact cell. The binding affinities for Ca^{2+} and other ions, and the free concentrations of ions in the cytoplasm, are not sufficiently well characterized for this to be possible. However, it seems likely that changes in the concentration of Ca^{2+} in the cytoplasm do lead to changes in the amount of calcium bound to calmodulin.

7C3 Effects of Ca^{2+} binding

Binding of Ca^{2+} to calmodulin induces marked conformational changes in the protein. There are four possible calcium/calmodulin complexes varying from one calcium bound to all four sites occupied. A number of studies have been done to compare the structure of calcium-free calmodulin and the $4Ca^{2+}$/calmodulin complex. The binding of four Ca^{2+} ions leads to a ten per cent increase in the content of alpha-helix. The hydrodynamic properties of the protein show little change suggesting that there is no gross change in protein shape. However, there are alterations in the sensitivity of particular amino acids to chemical modification, and changes in the sensitivity of particular peptide bonds to proteolytic cleavage, which suggests that the environment of many individual amino acids does change.

The four calcium-binding sites are all different. Examination of conformational change in the presence of different concentrations of Ca^{2+} indicates that the binding is an ordered process, Ca^{2+} binding to the four sites in a specific sequence. In purified calmodulin, the binding of one Ca^{2+} leads to a conformational change which enhances the binding of the second Ca^{2+}. There is relatively little additional conformational change on binding of the third and fourth Ca^{2+} ions. Thus the Ca^{2+} binding may vary from one to four over a relatively wide range of free Ca^{2+} concentration, and is probably cooperative. This raises the possibility that different enzymes may be affected by different calcium–calmodulin complexes. This would allow for wide variations in the sensitivity of different enzymes to changes in free Ca^{2+} concentration, while making use of a single basic mechanism (see below).

7D Interaction of calcium/calmodulin complexes with target proteins

In most cases, calmodulin only binds to a target protein in the presence of Ca^{2+} ions. Many of the enzymes which are affected by calmodulin have been purified and their structures described. Several are single subunit proteins with markedly differing structures. Calmodulin-sensitive enzymes do not therefore share a common calmodulin-binding subunit.

In a few cases, notably phosphorylase kinase, calmodulin is permanently bound to the enzyme and the association is not affected by the presence or absence of Ca^{2+} ions. In this case, calmodulin functions essentially as a Ca^{2+}-binding subunit of the enzyme. Phosphorylase kinase is the only well-characterized soluble protein which contains tightly bound calmodulin. Cell membranes, however, contain calmodulin which can only be removed by relatively severe physical treatments which disrupt the membrane structure extensively. It seems likely that this membrane-associated calmodulin is tightly bound to membrane proteins, but the particular proteins concerned have not been identified.

7D1 Concentration considerations

The interaction between calmodulin and most soluble calmodulin-sensitive proteins depends upon the presence of Ca^{2+}. It follows from this that the formation of the Ca^{2+}/calmodulin/enzyme complex depends upon the concentration of all three components, the dissociation constant of the calcium/calmodulin complex and the dissociation constant for the binding of calcium/calmodulin to the enzyme. The concentration of calmodulin in the cytoplasm is about 10^{-5} mol/l while the resting free Ca^{2+} is about 10^{-7} mol/l. The enzyme concentration is probably also much lower than the concentration of calmodulin. The relevant dissociation constant for the calcium/calmodulin complex is difficult to determine precisely since there is cooperativity between the four sites and it is uncertain how many sites have to be occupied to produce an effect. However, it is likely to be between 10^{-6} mol/l and 10^{-5} mol/l. The constant for the dissociation of calcium/calmodulin from the calcium/calmodulin/enzyme complex is about 10^{-8} mol/l, but at the concentrations present in the cell the effective dissociation constant is likely to be higher.

A number of conclusions follow from these values. The cytoplasmic free Ca^{2+} concentration varies between 10^{-7} and 10^{-6} mol/l which is likely to cause a change in the Ca^{2+} occupancy of calmodulin from one per cent to about ten per cent. Since there is a large excess of total calmodulin over total enzyme and the relevant dissociation constant is low, a change in Ca^{2+} occupancy over this range should produce a substantial change in the association of calmodulin wth calmodulin-sensitive enzymes. This is analogous to the situation with hormones where a low receptor occupany is required for the maximal response (see Section 2C). It seems likely that at a cytoplasmic Ca^{2+} concentration of 10^{-7} mol/l there will be no activation of enzymes by calmodulin while at a concentration of 10^{-5} mol/l all calmodulin-sensitive enzymes will be maximally activated. If the binding of Ca^{2+} to calmodulin was approximately hyperbolic a concentration of 10^{-6} mol/l would give about 50% activation. Since the binding is cooperative and may also be affected by the formation of a complex with the target enzyme the extent of activation at 10^{-6} mol/l may be much greater.

7D2 Effects of calmodulin binding to the target enzyme

The conformation of calmodulin is affected by the binding of calcium. The possibility exists that the conformation may also change as a result of the binding of the calcium/calmodulin complex to a target protein. This in turn may affect the binding affinity for Ca^{2+} ions. This situation always arises when one protein operates upon another and applies equally to the action of protein kinases or phosphatases. The interaction

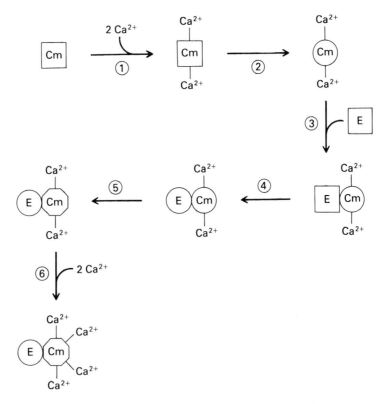

Fig. 7.3. Hypothetical sequence of interactions between Ca^{2+}, calmodulin and target enzyme. Cm=calmodulin, E=target enzyme. (1) Calmodulin binds 2 Ca^{2+} at the high-affinity sites, (2) conformational change in calmodulin, (3) calmodulin/2 Ca^{2+} complex binds to target enzyme, (4) conformational change in enzyme, (5) second change in calmodulin conformation consequent upon change in enzyme conformation. Increase in Ca^{2+} affinity of sites 3 and 4, (6) association of 2 further Ca^{2+}.

is dependent upon the activity of the modulator protein, in this case calmodulin, but the activity of the modulator can itself be affected by the target enzyme. The conformation of the target enzyme will change depending upon the binding of substrate or of other metabolites which affect its activity. This in turn may affect either the binding of calmodulin or the extent of conformational change in the calmodulin after binding, or both. In addition to this we have to consider the fact that calmodulin has four different binding sites for Ca^{2+}. Different calcium/calmodulin complexes may be required to modulate the activity of different enzymes.

Taking all this into consideration, it is clear that there are almost infinite possibilities for subtle variations in the functioning of the system. This can be illustrated by the hypothetical example shown in Fig. 7.3. In this case two calcium-binding sites need to be occupied to give the calmodulin conformation required for binding to the enzyme. The $(Ca^{2+})_2$/calmodulin complex binds to the enzyme and this initiates a conformational change in both the target enzyme and calmodulin. The conformational change in the target enzyme alters its activity while the conformational change in calmodulin increases the binding affinity of all the Ca^{2+} sites. As a result, sites 3 and 4 become occupied, stabilizing the conformational change of both calmodulin and the target enzyme. The binding of other effectors to the target enzyme could reverse or further enhance the increase in the Ca^{2+}-binding affinity of calmodulin.

Such an increase in Ca^{2+}-binding affinity on association of calmodulin with its target protein has important implications. The calcium will not all be released when the free Ca^{2+} drops back to the ground state, and the enzyme activation, or in other words the physiological response, may persist. This would allow a transitory increase in cytoplasmic Ca^{2+} to produce a prolonged change in enzyme activity. To reverse the change in the activity of the target enzyme, the free Ca^{2+} would have to fall below the original resting state. Alternatively, the binding of another effector to the target enzyme, or indeed to calmodulin, might reduce the affinity for Ca^{2+} so that Ca^{2+} is released and the whole activation process reversed. This is an important and interesting point. As yet there is relatively little direct evidence to resolve the many possibilities. Only one target enzyme, cyclic AMP phosphodiesterase, has been studied at all extensively. The binding of calmodulin to phosphodiesterase is dependent upon some of the Ca^{2+}-binding sites being occupied. The interaction also affects the affinity of the unoccupied sites for Ca^{2+}, and binding of the substrate cyclic 3',5'-AMP to phosphodiesterase affects the calcium-binding affinity of associated calmodulin. There is, therefore, some direct evidence for the type of interaction discussed above.

7E Processes affected by calmodulin

Processes known to be affected by calmodulin are listed in Table 7.1. The list is not exhaustive and new effects are reported each year. The following sections describe some of the better characterized calmodulin responses.

Table 7.1. Calmodulin-sensitive processes

	Tissue shown in
Cyclic AMP phosphodiesterase	Most
Adenylate cyclase	Brain
Plasma membrane Ca^{2+}-ATPase	Erythrocytes
Microtubule aggregation	Many
Myosin light chain kinase	Smooth muscle
Phosphorylase kinase	Smooth and skeletal muscle, liver
Other protein kinases	Various

7E1 Cyclic nucleotide phosphodiesterase

Calmodulin was first discovered in the early seventies as a result of studies upon a soluble Ca^{2+}-dependent cyclic nucleotide phosphodiesterase. The enzyme used both cyclic AMP and cyclic GMP as substrates. The K_m in the absence of Ca^{2+}/calmoldulin is high in both cases: 1.5 mmol/l for cyclic AMP and 0.26 mmol/l for cyclic GMP. Activation by the calcium/calmodulin complex causes a reduction in K_m for cyclic AMP to 0.1 mmol/l and for cyclic GMP to 0.02 mmol/l. In most cell types, cyclic AMP and cyclic GMP reach concentrations of about 10 μmol/l and 1 μmol/l, respectively, after stimulation. This means that the reduction in K_m will cause a very marked stimulation in the rate of breakdown of the cyclic nucleotides. The V_{max} for cyclic AMP also increases which will lead to a further increase in the rate of breakdown. The activation is instantaneous and reversed upon dilution; that is, it depends upon the concentration of the calcium/calmodulin complex. This suggests that activation is by a direct binding interaction involving phosphodiesterase and the Ca^{2+}/calmodulin complex. The binding stoichiometry appears to be 1:1, but the number of Ca^{2+}-binding sites which need to be occupied on calmodulin has not been clearly established. The measurement of this is complicated by the fact that the Ca^{2+}-binding affinity changes upon association of calmodulin with the phosphodiesterase. To determine whether or not the enzyme is likely to be controlled by Ca^{2+} under physiological conditions, we need to know the concentration of calmodulin, the concentration of phosphodiesterase, the number of Ca^{2+} sites which need to be occupied and their affinity, and the likely changes in free Ca^{2+} concentration. For many of these

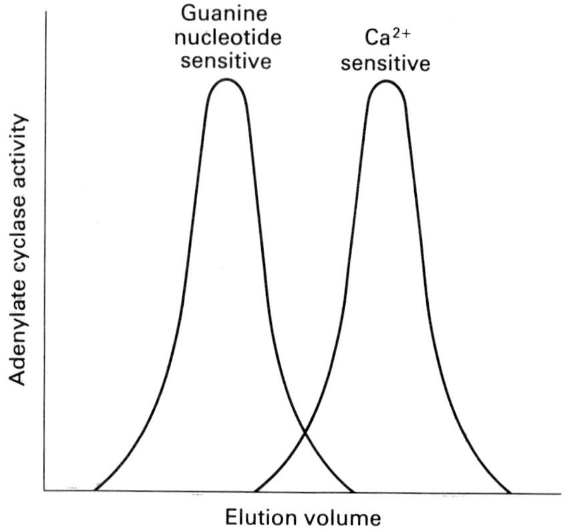

Fig. 7.4. Separation of Ca^{2+}- and guanine nucleotide-sensitive forms of brain adenylate cyclase by gel chromatography. The guanine nucleotide-sensitive form elutes first indicating a higher molecular weight, presumably reflecting the association of Gs (see Chapter 3).

factors only approximate figures are available, but it seems likely that the enzyme is regulated by changes in cytoplasmic free Ca^{2+} acting via calmodulin *in vivo*.

7E2 Adenylate cyclase

In most tissues, adenylate cyclase is inhibited by Ca^{2+}, the inhibition requiring fairly high concentrations. In the brain, however, it was found that micromolar concentrations of Ca^{2+} are needed for full activity. This was suggested by the observation that the Ca^{2+} chelator EGTA inhibits brain adenylate cyclase, in contrast to most other tissues where it tends, if anything, to activate. Later studies showed that the brain contained two distinct populations of adenylate cyclase. One was activated in the usual way by guanine nucleotides activating through a guanine nucleotide-binding protein. The other was insensitive to guanine nucleotides, but was activated by Ca^{2+} acting through calmodulin. In detergent solubilized preparations, the two fractions can be separated by gel filtration, the guanine nucleotide-sensitive fraction having a higher molecular weight (Fig. 7.4). They may also be separated by binding the calcium-sensitive form to a calmodulin affinity column. The existence of a calcium-sensitive form explains the observation that in brain, α_1-adrenergic agonists activate adenylate cyclase. The activation results from increase in cytoplasmic Ca^{2+} rather than a direct response to the receptor.

7E3 Ca^{2+} transport

Calmodulin has been shown to activate the Ca^{2+}-ATPase of erythrocyte plasma membranes. The activation is Ca^{2+} dependent and appears to be by a straightforward reversible association of the calcium/calmodulin complex with the pump. The apparent K_m for Ca^{2+} is about 0.8 μmol/l which is a little higher than the resting cytoplasmic free Ca^{2+} concentration. Similar results have been reported in synaptosome plasma membranes. The effect has not been demonstrated in other cell types but this may reflect the difficulty of measuring Ca^{2+} pump activity in cells other than erythrocytes. Activation of the Ca^{2+} pump by calmodulin would tend to reverse a rise in cytoplasmic free Ca^{2+}.

The three mechanisms involving calmodulin described above are all concerned with modulation of either the Ca^{2+} signal itself or of the cyclic AMP signal. They offer the possibility of self-regulation of the Ca^{2+} signal and of interaction between the two main messenger systems. In other cases, calcium acts directly upon cellular function through calmodulin.

7E4 Microtubules

Microtubules, formed by the association of tubulin, occur both in the cytoplasm and associated with the plasma membrane. They are a component of the cytoskeleton and are a major component of the mitotic spindle. They are believed to be essential for cell mobility and to be involved in secretory processes. All these processes involve Ca^{2+}, and it has been known for some time that microtubules disaggregate in the presence of Ca^{2+}. More recently it has been shown that the effects of Ca^{2+} are mediated by calmodulin.

7E5 Protein kinases

A number of protein kinases have been found to be activated by Ca^{2+} calmodulin complexes. In smooth muscle, the activation of myosin light chain kinase by Ca^{2+} is mediated by calmodulin. In skeletal muscle, the similar protein, troponin C, plays the same role.

Phosphorylase kinase *b* in smooth muscle is also activated by calcium via calmodulin. In this case the association of calmodulin with the enzyme does not depend upon the presence of Ca^{2+}, and calmodulin functions as a Ca^{2+}-binding subunit of the enzyme.

These protein kinase activations are well characterized with well-defined substrates. It is clear, however, that there are both membrane-bound and soluble calmodulin-dependent protein kinases of wider specificity. In the liver, vasopressin promotes the Ca^{2+}-dependent phosphorylation of ten soluble proteins, and calmodulin-dependent protein phosphorylation has been shown in many different plasma membrane preparations. The protein substrates for calmodulin-activated protein kinases are less well characterized than for the cyclic AMP-dependent protein kinases. In many cases, however, it seems that the same proteins may be phosphorylated by both mechanisms. The two kinases need not incorporate phosphate at the same site raising the possibility of their either reinforcing or antagonizing each others effects (see Chapter 8). In any case, it seems likely that protein phosphorylation provides the common mechanism by which cyclic AMP and Ca^{2+} ions produce their effects on enzyme activity.

7F Troponin C

Troponin C is the calcium-binding protein associated with contractile elements in cardiac muscle and skeletal muscle. It is highly acidic, and has a molecular weight of 12 000. The protein is very similar to calmodulin, containing four Ca^{2+}-binding sites: two with a binding affinity of 0.27 µmol/l and two with an affinity of 33 µmol/l. Comparison of the amino acid sequences shows 50% homology with calmodulin and a further 25% conservative substitutions, so the two proteins are clearly closely related. Unlike calmodulin, troponin C does not occur independently in solution, and only two target proteins have been identified. Its major role is clearly to mediate the activation by Ca^{2+} of myosin light chain kinase in skeletal and heart muscle. It has also been shown to activate skeletal muscle phosphorylase kinase *b* (see Section 14F).

Suggestions for further reading

Cheung W.Y. (Ed) (1980) Calmodulin. In *Calcium and Cell Function*, vol. 1. Academic Press.
Dedman J.R., Walsh M.J., Kaetzel M.A., Pardue R.L. & Brinkley B.R. (1982) Localisation of calmodulin in tissue culture cells. In *Calcium and Cell Function*, vol. 3, pp. 455–473. Academic Press.
Means A.R. (1982) Calmodulin, properties, intracellular location and multiple roles in cell regulation. *Recent Progress in Hormone Research*, **37**, 332–369.
Rasmussen H. (1983) *Calcium and Cyclic AMP as Synarchic Messengers*, chapter 3. Wiley.

Chapter 8 Interactions between cyclic AMP and Ca^{2+} as messengers

8A Introduction

As discussed in Chapters 3–7, cyclic AMP and Ca^{2+} function as the major messengers for a large number of hormones which promote rapid cellular responses. Many hormones make use of both messenger systems, the effects being mediated by a different class of receptors for each system. Cyclic AMP and Ca^{2+} may also be involved in the control of long-term responses and this will be discussed in Chapter 12. There are many instances in which one of the two messengers acts to either inhibit or reinforce the cellular response of the other messenger. There are also cases where both messenger systems modulate the same metabolic pathway in parallel, and one or two cases where the stimulation of a metabolic pathway requires the interaction of both Ca^{2+} ions and cyclic AMP. Many of these interactions have already been mentioned, but in this chapter the nature of interactions between the two hormone messenger systems will be dealt with more systematically.

8B Levels of interaction

There are four levels at which the two systems can interact (Fig. 8.1).
1 A rise in one messenger could enhance or inhibit the binding of hormone to the receptor working via the other messenger.
2 The metabolism of the other messenger could be affected. Cyclic AMP might enhance or inhibit either Ca^{2+} influx into the cytoplasm or Ca^{2+} efflux from the cytoplasm. Ca^{2+} ions might activate or inhibit either the synthesis of cyclic AMP by adenylate cyclase or its breakdown by phosphodiesterase.
3 Both messengers act through a protein—in the case of cyclic AMP, by binding to the regulatory subunit R of cyclic AMP-dependent protein kinase leading to activation of the kinase (Chapter 4), and, in the case of Ca^{2+} ions, by binding to calmodulin (Chapter 7). Thus the possibility exists that the other messenger might either enhance or inhibit this interaction. Cyclic AMP binding leads to the release and activation of the catalytic unit of the protein kinase, and many of the effects of calcium are thought to result from the activation of Ca/calmodulin-dependent protein kinase. In either case, the other messenger might directly effect the protein kinase activity.
4 The enzymes responsible for the physiological response may be affected. This can happen in a number of ways. The other messenger could directly bind to and affect the activity of the enzyme. It may affect the susceptibility of the enzyme to phosphorylation by the other messenger's protein kinase. The two messengers might promote phosphoryla-

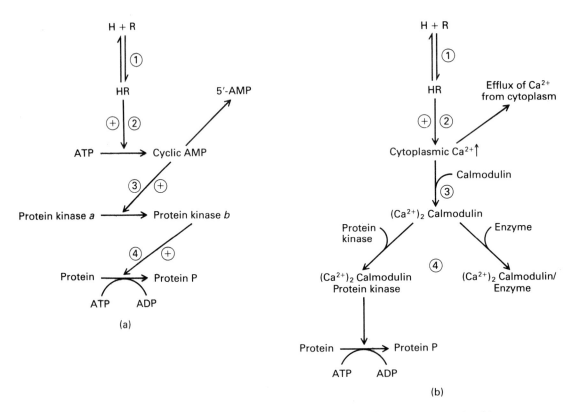

Fig. 8.1. Possible sites of interaction between the cyclic AMP and Ca^{2+} signals. (a) Cyclic AMP mediated. (b) Ca^{2+} mediated. See text for description of steps 1–4.

tion at different sites on the enzyme, with different effects on the activity.

From the above, it is apparent that there are a very large number of possible types of interaction. Examples can be found of many but by no means all of these. There is no evidence that cyclic AMP affects the binding of Ca^{2+}-mobilizing hormones to their receptors, or that increased cytoplasmic Ca^{2+} affects the receptor binding of hormones which activate adenylate cyclase.

There is also no evidence of direct effects of Ca^{2+} on cyclic AMP-dependent protein kinase, although there are many instances of effects of Ca^{2+} on proteins which are substrates of the protein kinase. Conversely, there is no evidence that cyclic AMP affects calmodulin either directly or by stimulating its phosphorylation. However, the response of enzymes to calmodulin may be affected by cyclic AMP-dependent phosphorylation.

8C Effects on the metabolism of the other messenger

8C1 Mechanisms for control of cyclic AMP levels by Ca^{2+}

There are three well-characterized effects of Ca^{2+} ions on the enzymes of cyclic AMP metabolism. In brain adenylate cyclase is activated by Ca^{2+} acting through calmodulin. The Ca^{2+}-sensitive cyclase appears to be one of two distinct populations of the enzyme, the other being controlled by guanine nucleotide-binding proteins in the usual way (see Chapters 3 and 7). In most tissues, adenylate cyclase is inhibited by Ca^{2+} ions. The concentration required for substantial inhibition is quite high and there is no evidence for any involvement of calmodulin. The effect probably reflects competition of Ca^{2+} with Mg^{2+} either for a Mg^{2+}-binding site, or as $CaATP^{2-}$, and is of doubtful physiological significance.

In most cells, cyclic AMP phosphodiesterase is activated by Ca^{2+} acting through calmodulin (see Chapter 7). Of the three effects this is most likely to be physiologically important.

8C2 Effects of cyclic AMP on Ca^{2+} metabolism

Many mechanisms have been proposed by which cyclic AMP may affect cytoplasmic free Ca^{2+} levels. It has been suggested to increase both Ca^{2+} influx and efflux across the plasma membrane, to increase Ca^{2+} uptake by the endoplasmic or sarcoplasmic reticulum, and to increase both Ca^{2+} uptake and release from the mitochondria. Many of these effects have been reported in a limited number of cell types and other workers have often found them difficult to repeat. There is no clear picture of the mechanisms by which cyclic AMP might affect cytoplasmic free Ca^{2+} levels. This is not surprising since the modulation of cytoplasmic free Ca^{2+} concentration is generally poorly described. However, there is good indirect evidence that cyclic AMP alters the cytoplasmic free Ca^{2+} concentration in a number of cell types. Some examples of effects of cyclic AMP and Ca^{2+} on each others metabolism are described in the next section.

8C3 Evidence for effects of cyclic AMP and Ca^{2+} on each others metabolism

Rat erythrocytes

The Ca^{2+} metabolism of erythrocytes is relatively simple since they have no mitochondria and little endoplasmic reticulum. The cytoplasmic Ca^{2+} concentration is therefore controlled by the plasma membrane carriers. There is only one efflux carrier, the Ca^{2+}-ATPase, since the Na^+/Ca^{2+} exchanger is absent. The erythrocytes of many species have very low adenylate cyclase activity. In the rat, however, there is an adenylate cyclase which is activated by β-adrenergic agonists.

The change in cyclic AMP in rat red blood cells in response to activation by adrenaline is shown in Fig. 8.2. The concentration of cyclic AMP increased sharply, fell over a period of about 20 minutes and then increased again to reach a maximum after about one hour. If the cells were preincubated with EGTA to deplete them of Ca^{2+}, the dip in cyclic AMP was abolished. If Ca^{2+} was restored to the depleted cells by the addition of Ca^{2+} together with the ionophore A23187, to render the plasma membrane permeable to Ca^{2+} ions, the cyclic AMP level fell rapidly. The change in cyclic AMP can be considered to occur in three phases (Fig. 8.2). An estimate of the adenylate cyclase activity can be obtained from the rate of rise of cyclic AMP concentration after addition of an inhibitor of phosphodiesterase. There appeared to be little difference between the adenylate cyclase activity in phase 2 and phase 3 and the activity was similar to the activity in fresh cells immediately after the addition of adrenaline. It seems unlikely that changes in adenylate cyclase activity provide an explanation for the fall in cyclic AMP. However, the phosphodiesterase activity did change. The activity in plasma

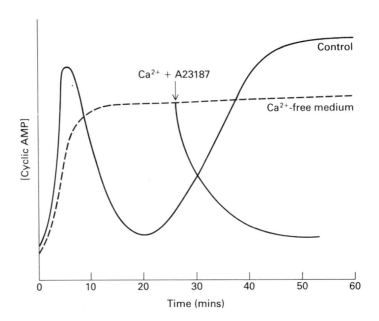

Fig. 8.2. Effect of adrenaline on cyclic AMP levels in rat erythrocytes. Adrenaline was added at 0 time and was present throughout the incubation.

membranes isolated from cells in phase 2 was substantially higher than the activity in membranes from phase 3 cells. Finally, during the third phase the plasma membrane Ca^{2+}-ATPase activity increased. The observations suggest the following sequence of events.

1   The initial rise in cyclic AMP causes a rise in cytoplasmic free Ca^{2+} by a mechanism which has not as yet been characterized.

2   The rise in Ca^{2+} causes an increased concentration of the Ca^{2+}/calmodulin complex, which activates both cyclic AMP phosphodiesterase, leading to a fall in cyclic AMP, and the Ca^{2+}-ATPase, leading to a fall in cytoplasmic Ca^{2+}. The fall in cytoplasmic Ca^{2+} reverses these effects leading to the second rise in cyclic AMP.

Isolated rat liver cells

Changes in cytoplasmic Ca^{2+} concentration in isolated liver cells have been studied extensively, using a wide variety of methods (Chapter 6). Such studies have tended to concentrate on the effects of Ca^{2+}-mobilizing hormones and relatively little attention has been given to the relationship between cyclic AMP concentration and cytoplasmic Ca^{2+} concentration. However, glucagon increases the cytoplasmic Ca^{2+} concentration in rat liver cells. Recently the use of the Ca^{2+} indicator, Quin 2, has allowed a comparison between the time course of the rise in Ca^{2+} in response to Ca^{2+}-mobilizing hormones and the increase in Ca^{2+} in response to glucagon. The increase in response to glucagon is much slower, being detectable after 5 seconds and maximal after 20 seconds, while adrenaline gives a detectable increase in Ca^{2+} after 1 second and the effect is maximal after 5 seconds (Fig. 8.3). It seems likely that the

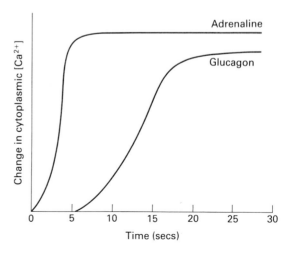

Fig. 8.3. Time course of change in cytoplasmic Ca^{2+} in rat liver cells in response to adrenaline and glucagon. Both hormones were added at 0 time and were present throughout the incubation. Note the short time course, hence no sign of a subsequent fall in Ca^{2+}.

effect of glucagon on cytoplasmic Ca^{2+} levels is secondary to a rise in the concentration of cyclic AMP. The details of the mechanisms involved have not been characterized.

Heart muscle

It has been known for many years that in heart muscle, β-adrenergic agonists increase both the rate and amplitude of contraction and the rate of relaxation. The effect is caused by a rise in cyclic AMP and can be reproduced by exogenous cyclic AMP or analogues of cyclic AMP. Therefore, cyclic AMP must either affect the changes in Ca^{2+} concentration during muscle contraction or modulate the response of the contractile apparatus to Ca^{2+} ions. It has been suggested that the rate of entry of Ca^{2+} across the cell membrane through the voltage dependent Ca^{2+} channels is increased (see Chapter 5). Cyclic AMP-dependent phosphorylation of the membrane has also been demonstrated, but as yet no causal relationship has been established between the two effects. There is better evidence for an effect of cyclic AMP on Ca^{2+} uptake by the sarcoplasmic reticulum. Both the rate and extent of accumulation are increased. The rate of accumulation appears to be controlled, at least in part, by a protein called phospholamban which is associated with the sarcoplasmic reticulum Ca^{2+}-pumping ATPase. Phospholamban is phosphorylated on a serine residue by a Ca^{2+}/calmodulin-dependent protein kinase and this activates Ca^{2+} uptake. There is also a cyclic AMP-dependent phosphorylation on a different serine residue, which has no effect on Ca^{2+} uptake by itself but potentiates the effect of the Ca^{2+}-dependent phosphorylation. Thus, in this case, cyclic AMP is working on the Ca^{2+} signal at two levels. It enhances the effect of Ca^{2+} and also causes a change in Ca^{2+} metabolism. The consequence of the increased rate of uptake is an increased rate of relaxation. The consequence of the increased extent of uptake is an increase in the amount of Ca^{2+} released leading to a stronger contraction.

8D Interactions at the level of intermediary metabolism

There are many possible consequences of cyclic AMP and Ca^{2+} acting upon the same pathway or the same enzyme. The effects may be redundant, in that either regulator will maximally activate the pathway. Thus in the presence of maximally effective concentrations of one regulator, the other will have no effect. The effect of one regulator may be either enhanced or diminished by the presence of the other. Both regulators may be necessary to produce an effect. Finally, the two regulators may cause the same physiological response acting via completely separate pathways. There are examples of most of these situations and some of these are described below.

8D1 Effects of Ca^{2+} and cyclic AMP on liver metabolism

Glucose release from liver cells is rapidly activated by either increased cytoplasmic Ca^{2+} or by an increase in cyclic AMP levels. Two pathways are affected: glycogenolysis and gluconeogenesis.

The effect of hormones on the phosphorylation state of proteins in a cell can be examined by incubating the cell in the presence of $[^{32}P]$-P_i until the label equilibrates into the γ-phosphate of ATP, adding the hormone, separating the proteins by SDS gel electrophoresis and comparing the labelled bands with the control. When this is done with liver cells, glucagon increases the phosphorylation of twelve proteins. Vasopressin or adrenaline increase the phosphorylation of eleven of the same proteins. It seems likely then that the two signals act in a redundant fashion. That is, glucagon produces its effect by phosphorylation through the cyclic AMP-dependent protein kinase, while the Ca^{2+}-mobilizing hormones act on the same proteins through a Ca^{2+}/calmodulin-dependent protein kinase. However, the situation may not be quite so straightforward. The single protein which is phosphorylated in response to glucagon, but not vasopressin, can be identified as phosphorylase kinase. Phosphorylase kinase has an absolute requirement for Ca^{2+} ions acting via the δ subunit which is calmodulin (see Chapter 14). The *b* form of phosphorylase kinase may be activated by one of two different mechanisms. Phosphorylation of the β subunit by the cyclic AMP-dependent protein kinase converts the enzyme to the *a* form, which is still Ca^{2+} dependent but requires a much lower concentration of Ca^{2+}. Alternatively, phophorylase kinase *b* may be activated by a rise in the cytoplasmic Ca^{2+} acting via the δ' subunit which may be either soluble calmodulin or troponin C. In this case, the two mechanisms act upon the same enzyme to produce the same end result—the activation of phosphorylase (see Section 14E).

Many of the eleven enzymes which are phosphorylated in response to both hormones have been identified (Table 8.1). In the case of phosphorylase, the phosphorylation takes place at the same site and by the same enzyme, phosphorylase kinase. The two signals operate via a common pathway. In the case of glycogen synthetase, different sites are phosphorylated by the cyclic AMP-dependent protein kinase and phosphorylase kinase. Both phosphorylations inhibit the enzyme, but the phosphorylation of both sites leads to a more intense inhibition than the

Table 8.1. Enzymes phosphorylated in liver cells in response to vasopressin and glucagon

---
Phosphorylase

Glycogen synthetase

Pyruvate kinase

Phosphofructokinase 2/
  fructose bisphosphatase 2

Acetyl CoA carboxylase

---

phosphorylation of one site or the other individually (see Chapter 14F). Thus there are two effects which reinforce each other—one dependent upon cyclic AMP mediated by cyclic AMP-dependent protein kinase, and one dependent upon Ca^{2+} ions and cyclic AMP mediated by phosphorylase kinase.

It is not known if the other enzymes which are phosphorylated in response to both signals have phosphate incorporated at the same site or at different sites. Phosphorylation at separate distinct sites allows the two signals to reinforce each other, oppose each other or simply to act in parallel. In general, it appears that in liver, cyclic AMP and Ca^{2+} ions act upon the same enzyme by a similar mechanism to produce the same physiological response. To this extent the effects are redundant, but at the level of the individual enzymes more complex interactions may occur.

8D2 Antagonistic effects

There are many cases where cyclic AMP and Ca^{2+} ions tend to produce opposite effects on metabolism and to antagonize the response to the other signal. Two examples are described below.

The platelet release reaction

The physiological role of platelets is to respond to injury. One aspect of the response is to release a number of different components from the cell. Among these are ADP, 5-hydroxytryptamine (5HT), fibrinogen, and platelet factor 4; ADP and 5HT promote platelet aggregation, and fibrinogen and platelet factor 4 promote blood clotting. At the same time prostaglandin biosynthesis is activated. The different components are released simultaneously and the release reaction is promoted by a number of external agonists including ADP, 5HT, thrombin, collagen, adrenaline and arachidonic acid. Release is also activated by the Ca^{2+} ionophore A23187, and it is clear that the signal for the release reaction is an increase in cytoplasmic free Ca^{2+}. Prostaglandin E_1 activates adenylate cyclase in platelets and inhibits the release reaction. The effect is enhanced by phosphodiesterase inhibitors and can be reproduced by the addition of exogenous cyclic AMP indicating that it is cyclic AMP dependent. Prostacyclin has the same effect but at much lower concentrations. Prostacyclin is synthesized in platelets at the same time as the release reaction and is probably the physiologically relevant agonist. It will thus have a damping effect on the release response. Adrenaline and ADP, which promote the release reaction primarily through the Ca^{2+} signal, also inhibit platelet adenylate cyclase. Adrenaline inhibits via α_2-adrenergic receptors and Gi (see Chapter 3). The mechanism of the ADP effect is poorly characterized.

It is obvious that cyclic AMP acts to antagonize the effect of Ca^{2+} on the release reaction, and that many of the agonists involved act in a concerted fashion on both signals. The mechanism by which cyclic AMP inhibits the release reaction is less clear. The effect could either be to alter the intracellular Ca^{2+} concentration or to affect the response to Ca^{2+}.

Smooth muscle

In smooth muscle, β-adrenergic agonists usually cause relaxation and there is good evidence that this effect is mediated by a rise in cyclic AMP. Cyclic AMP must either cause a reduction in cytoplasmic free Ca^{2+} or render the contractile apparatus less responsive to Ca^{2+}. There is evidence for three different mechanisms (Fig. 8.4).

(a) In isolated smooth muscle cells from toad stomach, β-adrenergic agonists cause a marked increase in K^+ influx and Na^+ efflux, associated with a rise in the concentration of cyclic AMP and an increase in the activity of cyclic AMP-dependent protein kinase (Fig. 8.4a). The effect is associated with an activation of the Na^+/K^+-ATPase. When isolated plasma membranes were incubated with ATP, cyclic AMP and cyclic AMP-dependent protein kinase, the activity of the Na^+/K^+-ATPase increased. It appears that the effect results from cyclic AMP-dependent phosphorylation of the plasma membrane, but it is not known if the phosphorylation is on the ATPase itself.

An increase in the Na^+ gradient across the plasma membrane, resulting from activation of the Na^+/K^+-ATPase, would be expected to result in increased efflux of Ca^{2+} on the Na^+/Ca^{2+} exchange carrier (see

(a)
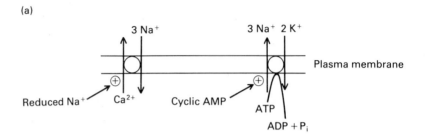

(b)
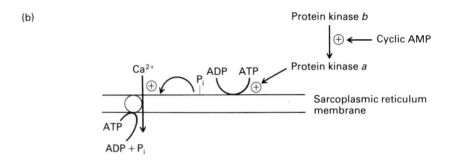

(c)
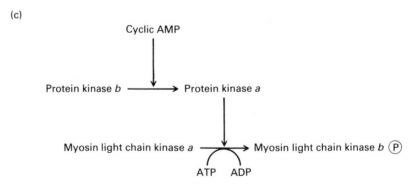

Fig. 8.4. Mechanisms of action of cyclic AMP in inhibiting smooth muscle contraction. (a) Activation of Na^+ efflux. (b) Activation of Ca^{2+} uptake in sarcoplasmic reticulum. (c) Phosphorylation of myosin light chain kinase. (a) – (c) correspond to points in the text.

Chapter 5). This implies that the Ca^{2+} concentration gradient across the plasma membrane is in equilibrium with the Na^+ gradient. There is some evidence that this is the case in this cell preparation, since reducing the external Na^+ concentration, and hence the Na^+ gradient, led to muscle contraction, implying an increase in cytoplasmic Ca^{2+} concentration.

(b) In smooth muscle cells from a number of sources, the sarcoplasmic reticulum is phosphorylated by cyclic AMP-dependent protein kin-

127 Interactions between cyclic AMP and Ca^{2+} as messengers

ase (Fig. 8.4b). This leads to an activation of Ca^{2+} uptake which could cause relaxation of the muscle in response to a rise in cyclic AMP. The mechanism is similar to the effects of β-adrenergic agonists in the heart. (c) Both the effects of cyclic AMP described above operate by altering the metabolism of Ca^{2+}. The third mechanism involves a modulation of the response to Ca^{2+}. Cyclic AMP-dependent protein kinase will phosphorylate myosin light chain kinase reducing its activity (Fig. 8.4c). The effect is to reduce the sensitivity of the myosin light chain kinase to activation by Ca^{2+}/calmodulin, inhibiting the Ca^{2+}-dependent phosphorylation of myosin light chain. This tends to relax the muscle.

These effects have been reported in smooth muscle from a number of different species. They do not necessarily all occur in all types of smooth muscle.

8D3 Parallel effects: the blow fly salivary gland

The blow fly salivary gland secretes saliva in response to 5-hydroxytryptamine. There is a net transport of KCl across the epithelial cells of the gland leading to changes in osmotic pressure which result in an associated movement of water (Fig. 8.5). The transport of KCl involves active transport of K^+ ions and a passive flux of Cl^- ions.

5HT increases both cyclic AMP and cytoplasmic free Ca^{2+}. Either signal will cause some response, but a maximal response of prolonged duration requires both cyclic AMP and Ca^{2+} acting together. Early studies suggested that the effect of 5HT on secretion was mediated through cyclic AMP. All the components of a cyclic AMP-dependent system are present. The agonist raised cyclic AMP concentration in the cell, and the addition of exogenous cyclic AMP induced secretion. Later it became apparent that this was only part of the story.

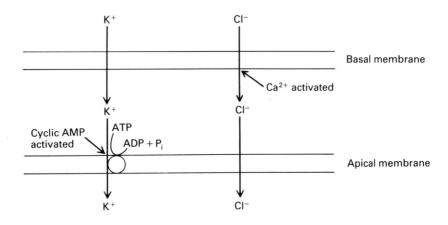

Fig. 8.5. Transport of KCl across insect salivary gland epithelial cell. The process operates essentially as two independent pathways: one activated by cyclic AMP and the other by Ca^{2+}.

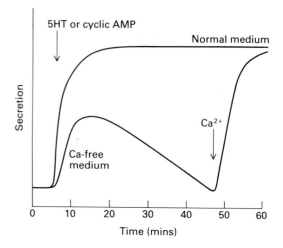

Fig. 8.6. Ca^{2+} requirement for secretion in insect salivary gland. The response to either 5HT or exogenous cyclic 3′,5′-AMP declines in Ca^{2+}-free medium but may be restored by the addition of Ca^{2+}.

In Ca^{2+}-free medium, the response to either 5HT or exogenous cyclic AMP was reduced in both extent and duration (Fig. 8.6) suggesting that there is also a Ca^{2+}-dependent component in the response to 5HT. If Ca^{2+} was added back, secretion was restored. 5HT was found to stimulate calcium influx and also the appearance of calcium in the saliva. The calcium ionophore A23187 mimicked the action of 5HT, stimulating secretion. However, the effects were slower, and, in the case of the stimulation of secretion, smaller than the effects of the hormone. It is clear that 5HT affects both cyclic AMP and Ca^{2+} levels and that either messenger can activate the secretory response. However, both messengers are needed for a maximal effect of the hormone. This raises two questions.

1 Do cyclic AMP and Ca^{2+} act by a common mechanism, or by different mechanisms?
2 Do cyclic AMP and Ca^{2+} affect each other's metabolism?

The two signals act by different mechanisms. Cyclic AMP increases active K^+ transport in the lumen membrane, while Ca^{2+} induces a marked increase in the passive Cl^- flux (Fig. 8.5). Since the final response involves the transport of KCl, the two effects are complementary.

The two signals do affect each other's metabolism. The Ca^{2+} ionophore A23187, while activating secretion, causes a fall in the concentration of cyclic AMP. This suggests that Ca^{2+} either inhibits adenylate cyclase or, more probably, activates phosphodiesterase. This conclusion is supported by the observation that 5HT causes a larger rise in cyclic AMP in Ca-depleted cells than in control cells. Cyclic AMP appears to mobilize Ca^{2+} from an intracellular source, but this is poorly characterized.

8E Conclusions

The two major signal molecules which mediate the rapid intracellular effects of external agonists interact at many different levels. The interaction may be either positive or negative. This means that, using only two basic mechanisms, a hormone can produce a very large variety of responses in different cell types. This will depend partly upon the enzymatic profile of the cell, and partly upon whether the hormone has receptors which work through adenylate cyclase or the mobilization of Ca^{2+}, or both. A third very important factor will be the nature of the interaction between the two signals in the particular cell type. Taking all these factors into account, it becomes obvious that an almost infinite variety of responses can be generated by the same hormone acting on different tissues.

Suggestions for further reading

Berridge M.J. (1981) Phosphatidylinositol hydrolysis and calcium signalling. *Advances In Cyclic Nucleotide Research*, vol. 14, pp. 289–299.

Rasmussen H. (1983) *Calcium and cyclic AMP as synarchic messengers.* Wiley.

Chapter 9 Hormone action and phosphatidylinositol turnover

9A Introduction

The stimulation of phosphatidylinositol metabolism by an external agonist was first reported in the early 1950s, when it was shown that acetylcholine stimulated the incorporation of $[^{32}P]$-P_i into phosphatidylinositol in pancreas slices. Since then, many hormones and neurotransmitters have been shown to affect phosphatidylinositol metabolism in many different tissues (Table 9.1). In most cases it has been established that the agonist increases phosphatidylinositol breakdown, and that increased incorporation of labelled phosphate or inositol results from resynthesis. These effects have two things in common. First, the receptor involved does not stimulate or inhibit adenylate cyclase, although the agonist may affect adenylate cyclase activity via a different class of receptors. Second, in all cases which have been examined, agonists which increase phosphatidylinositol breakdown also increase the cytoplasmic free calcium concentration. This suggests that changes in Ca^{2+} and phosphatidylinositol metabolism may be related. The nature of this possible relationship is the major topic of interest in this area.

Table 9.1. Agonists which increase inositol phospholipid breakdown

Agonist	Tissue
Vasopressin	Brain, liver
Angiotensin	Liver
Adrenaline α_1	Fat, liver, brain
5-Hydroxytryptamine	Insect salivary gland
Concanavalin A	Lymphocytes
Nerve growth factor	Nerve cells
Epidermal growth factor	Fibroblasts
Acetylcholine (muscarinic)	Brain
Histamine	Brain
Thrombin	Platelets
Platelet activating factor	Platelets
Opiates	Brain

9B Cellular location of phosphatidylinositol

Phosphatidylinositol is a relatively minor component of the cell, phos-

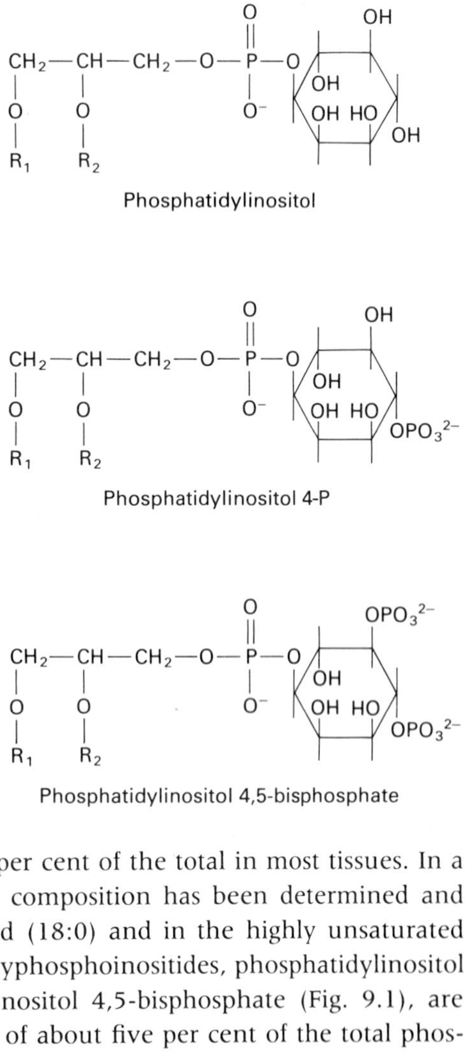

Fig. 9.1. Structures of inositol phospholipids. R_1 and R_2 represent fatty acids, commonly stearic and arachidonic.

pholipid comprising about five per cent of the total in most tissues. In a number of cases the fatty acid composition has been determined and shown to be rich in stearic acid (18:0) and in the highly unsaturated arachidonic acid (20:4). The polyphosphoinositides, phosphatidylinositol 4-phosphate and phosphatidylinositol 4,5-bisphosphate (Fig. 9.1), are much less abundant, consisting of about five per cent of the total phosphatidylinositol pool. Phosphatidylinositol occurs in all the membrane fractions of the cell but there is evidence that the polyphosphoinositides are localized in the plasma membrane. Cells where the plasma membrane represents a large proportion of the total cell membrane, for example myelin, tend to have relatively high contents of polyphosphoinositides. On the other hand, large cells where the plasma membrane is a relatively small proportion of the total membrane in the cell are relatively poor in polyphosphoinositides. Other membranes which have a functional relationship with the plasma membrane, for example the membranes surrounding secretory granules, have also been found to contain polyphosphoinositides.

9C Reactions of phosphatidylinositol metabolism

Having discovered that phosphatidylinositol metabolism was stimulated by hormones, an obvious first step was to attempt to characterize the enzymes involved. The reactions of phosphatidylinositol synthesis and breakdown form a cycle which is shown in Fig. 9.2. Breakdown to inositol 1-phosphate and diacylglycerol is catalysed by a phosphodiesterase sometimes called a phospholipase C. Synthesis requires the activation of diacylglycerol to phosphatidic acid and then to CDP-diacylglycerol, which will react with free inositol to give phosphatidylinositol.

The polyphosphoinositides, phosphatidylinositol 4-phosphate and phosphatidylinositol 4,5-bisphosphate, are formed from phosphatidylinositol by phosphorylation by ATP-dependent kinases, and the 4 and 5 phosphates can be removed by phosphomonoesterases. They may also be broken down by a phosphodiesterase to give inositol 1,4-bisphosphate or inositol 1,4,5,-trisphosphate. These may in turn be converted sequentially to inositol 1-phosphate and then free inositol by phosphomonoesterases. The reverse pathway from free inositol to inositol 1,4,5-trisphosphate does not appear to exist (Fig. 9.2).

The pathway is quite complicated and it has proved to be difficult to

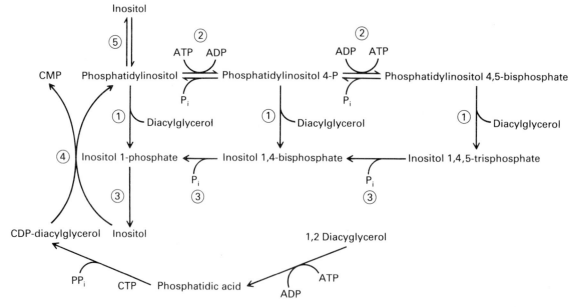

Fig. 9.2. The cycle of inositol phospholipid turnover. The types of reaction involved: (1) phosphodiesterases, (2) kinase, (3) phosphomonoesterases, (4) CDP-diacylglycerol inositol transferase, (5) phosphatidylinositol/inositol exchange.

characterize the enzymes involved in detail. A large part of the problem arises from the fact that many of the substrates involved are water-insoluble membrane components. These have to be presented to the enzymes in the form of membrane vesicles, and it is impossible to be sure that this adequately mimics the situation in the cell. The cellular location of the enzymes is also difficult to determine, since enzymes which may be in the soluble fraction in the intact cell tend to bind to membranes containing their substrate during cell fractionation.

9D Enzymes of inositol lipid metabolism

9D1 Phosphodiesterases

The effect of agonist action on phosphatidylinositol metabolism appears to be to stimulate breakdown (see Section 9E2). Accordingly, there has been a lot of interest in the enzymes which catalyse the initial breakdown reaction. Phosphatidylinositol phosphodiesterase has been extensively studied in liver where activity is found in the lysosomes and in the cytoplasm. The lysosomal enzyme is probably not involved in the hormonal stimulation of breakdown. The cytoplasmic enzyme has proved to be very difficult to characterize. Purification has given several different active fractions. Some of these undoubtedly result from

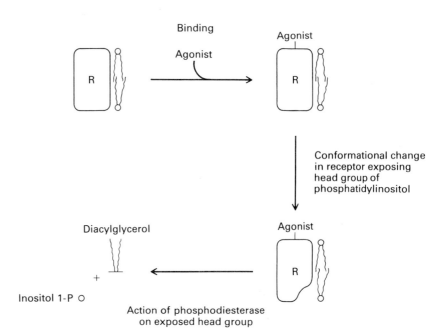

Fig. 9.3. Postulated effect of agonist binding on phosphatidylinositol breakdown.

proteolysis during the extraction and it is not known which fraction, or fractions, represents the physiological state of the enzyme. It is clear, however, that the enzyme is calcium dependent and located in the cytoplasm. As might be expected of a soluble enzyme working upon a substrate which is a membrane constituent, the form in which the substrate is presented to the enzyme is critical. In particular, other lipids present in the vesicles containing phosphatidylinositol used as substrate have a marked effect on the activity. This raises the possibility that the activity of the enzyme *in vivo* might be altered by changes in the orientation of the substrate lipids in the membrane, which might be affected by agoinst binding (see Fig. 9.3). The phosphodiesterase has been reported to be activated by arachidonic acid, which is one of the products of breakdown.

A single phosphodiesterase catalyses the breakdown of both phosphatidylinositol 4-phosphate and phosphatidylinositol 4,5-bisphosphate. This enzyme requires low levels of Ca^{2+} for activity. Higher Ca^{2+} concentrations enhance the activity further. However, other positively charged ions can do the same, suggesting that this effect is due to the neutralization of the negative charge on the substrate phospholipid. The enzyme is found in both soluble and membrane-bound fractions.

9D2 Phosphomonoesterases

It is not clear whether separate enzymes or a single enzyme catalyse the removal of phosphate from phosphatidylinositol 4-phosphate and phosphatidylinositol 4,5-bisphosphate. The activity occurs in both soluble and membrane-bound forms, and is Ca^{2+} dependent. Neutralizing the negative charge of the substrate with higher levels of Ca^{2+} or other charged species again enhances activity. It is not clear which form of the enzyme functions *in vivo*.

There are also phosphomonoesterases which will remove phosphate from the breakdown products inositol 1,4,5-trisphosphate, inositol 1,4,-bisphosphate and inositol 1-phosphate. A membrane activity specific for inositol 1,4,5-trisphosphate has been described in red blood cells. This might be an interesting enzyme since it has been suggested that the initial hormone response is the breakdown of phosphatidylinositol 4,5,-bisphosphate. The enzyme has not, however, been described in a hormone-responsive tissue.

9D3 Phosphatidylinositol kinase

The kinase which catalyses the phosphorylation of phosphatidylinositol to phosphatidylinositol 4-phosphate has been identified in a number of tissues. In liver, the enzyme is located in the plasma membrane, and it is reported to be inhibited by a rise in cyclic AMP levels in the cell, presumably as a result of cyclic AMP-dependent phosphorylation. The kinase which adds the final phosphate to the 5 position appears to be soluble.

9D4 CDP-Diglyceride inositol transferase

This enzyme is membrane bound, the highest specific activity being found in the microsomal fraction. The enzyme is inhibited by unsaturated fatty acids, but the physiological significance of this, if any, is not known.

The enzymology of phosphatidylinositol breakdown and synthesis is poorly understood. However, some conclusions are possible. All the enzymes involved in breakdown have a requirement for calcium. With one exception, the enzymes are not localized in the plasma membrane, and as yet there is no obvious candidate for an enzyme which may be directly activated by an occupied receptor in a manner analogous to the action of hormones on adenylate cyclase.

9E Effects of hormones on the metabolism of inositol phospholipids

Over the years, more and more information has accumulated about the effects of hormones on phosphatidylinositol metabolism. In spite of this, a real understanding of the mechanisms involved has remained elusive, and the physiological significance of the response has only recently become apparent.

9E1 Technical problems

The earliest observations showed an increase in the incorporation of [^{32}P]-inorganic phosphate into phosphatidylinositol in response to a number of different hormones in a number of different tissues. This type of experiment is both technically complicated and difficult to interpret. Incubating whole cells with [^{32}P]-P$_i$ results in the incorporation of label into all the phospholipids. Thus the phosphatidylinositol must be separated, not only from the very large number of other phosphate-containing compounds in the cell, but from all the other phospholipids. It is a minor component of the total phospholipid and the separation is tedious and time consuming. More serious problems can arise from failure to label the ATP pool to a steady state with ^{32}P. Hormones almost always alter the rate of turnover of ATP in cells. If the label in the γ-phosphate of ATP has not reached equilibrium with the specific activity of the free P_i pool, then the addition of the hormone may increase the label in the ATP. This may lead to increased labelling of phospholipids with no net increase in the amount of lipid present.

Later studies made use of [^{3}H]-inositol. This has the obvious advantage that only inositol-containing lipids will be labelled. However, it usually takes too long to label the cells to the point where the specific activity of phosphatidylinositol is the same as that of the free inositol. It is also difficult to wash the free labelled inositol away. This means that labelled inositol incorporation measurements, or measurements of release from prelabelled cells, is always affected by both synthesis and breakdown rates, although it is possible to choose conditions which maximize the effects of one pathway or the other. The exchange enzyme also allows the possibility of increase or decrease in phosphatidylinositol labelling with no change in net flux.

Glycerol labelling has been used, which avoids the exchange problem. However, all the phospholipids are labelled. There is also the possibility that the specific activity of glycerol phosphate can be altered by changes in glycolytic or gluconeogenic flux, both of which are liable to be affected by hormones.

Perhaps the method least open to problems of interpretation is to isolate the inositol lipids and to determine the amount present by determining the phosphorus content by chemical analysis. This method is very tedious indeed, and the sensitivity is poor so that large amounts of tissue are needed. However, the interpretation is unequivocal. The main problem is that it is unsuitable for measuring rapid time courses of response.

9E2 Evidence for hormone stimulation of breakdown

As described above, the early observations reported an increase in the incorporation of [^{32}P]-P_i. Due to the cyclic nature of pathways of phosphatidylinositol metabolism, this could result either from a stimulation of synthesis or from a stimulation of breakdown followed by resynthesis. To resolve this question, the effects of hormones on the total phosphatidylinositol content have been measured in a number of cell types. The total phosphatidylinositol content decreased, indicating that the main effect of the hormones was to increase the rate of breakdown of phosphatidylinositol.

Recently, better methods for measuring breakdown have been devised. The simplest approach is to preload the cells with [^{3}H]-inositol so that the phosphatidylinositol and polyphosphoinositides become labelled. The rate of release of inositol phosphates can then be measured in the presence and absence of the hormone, as inositol phosphates are easily separated from inositol by virtue of their charge. The main problem with this approach is that inositol phosphates are rapidly converted to inositol by the phosphomonoesterases (see Fig. 9.2). However, it has been found that inositol 1-phosphate monoesterase is strongly inhibited by lithium. Thus, when cells are incubated in the presence of lithium, the accumulation of inositol 1-phosphate gives a measure of the total breakdown of phosphatidylinositol and the polyphosphoinositides. It is not possible to determine which inositol phospholipid is broken down by phosphodiesterase activity since inositol 1,4,5-trisphosphate and inositol 1,4-bisphosphate will still be rapidly converted to inositol 1-phosphate. This method confirms the suggestion that hormones stimulate breakdown, and allows the effect to be measured more readily and accurately.

9E3 Pathways of breakdown

The three inositol phospholipids in the membrane, phosphatidylinositol, phosphatidylinositol 4-phosphate and phosphatidylinositol 4,5-bisphosphate, are readily interconvertible by the actions of phosphatidylinositol kinases and phosphomonoesterases. Thus any one of three inositol phospholipids could be the initial substrate when breakdown is stimulated by the hormone. The polyphosphoinositides are minor components, but either could be the initial substrate for breakdown and then be replenished from the much larger phosphatidylinositol pool, which, in the absence of increased synthesis, would slowly decrease (see Fig. 9.2).

This question can only be resolved by examining very rapid time courses of the appearance of inositol phosphates, in the hope of detecting an increase in inositol 1,4-bisphosphate or inositol 1,4,5-trisphosphate before the monoesterases convert them to inositol 1-phosphate. This has been done using the blow fly salivary gland which has an unusually large stimulation of phospholipid metabolism in response to 5-hydroxytryptamine. In this case it appears that there is first an increase in inositol 1,4,5-trisphosphate, followed by an increase in inositol 1,4-bisphosphate, and finally, a slow sustained increase in inositol 1-phosphate (see Fig. 9.4). Thus it appears that in this cell type, at least the initial substrate for breakdown is phosphatidylinositol 4,5-bisphosphate and the first product inositol 1,4,5-trisphosphate. It has been suggested that this is a general phenomenon, and in other cell types there is good

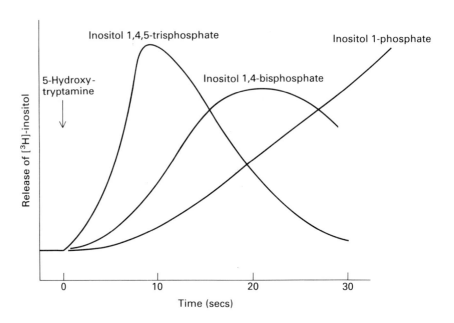

Fig. 9.4. Time course of release of inositol phosphates in insect salivary gland.

evidence for a more rapid breakdown of polyphosphoinositides than phosphatidylinositol. However, in most cases it is not obvious that phosphatidylinositol 4,5-bisphosphate is broken down any faster than phosphatidylinositol 4-phosphate.

9F Role of phosphatidylinositol breakdown in mediating the hormone response

It is clear that hormones stimulate phosphatidylinositol breakdown, although the mechanisms involved are incompletely characterized. The next question to consider is whether or not the stimulation of phosphatidylinositol breakdown is essential for the physiological response of the cell to the hormone. This has proved to be quite difficult to establish. The simplest approach would be to inhibit specifically the breakdown and then to see if the same inhibitor also blocked the response to the hormone. Unfortunately, no specific inhibitors of phosphatidylinositol breakdown have been discovered.

Perhaps the best evidence for a key role for the breakdown process comes from desensitization studies. If cells are exposed to a massive dose of a hormone which stimulates phosphatidylinositol breakdown, they become desensitized over a period of half an hour or so and the response to the hormone declines (Fig. 9.5). Desensitization is common to the action of most extracellular agonists. In this case the loss of res-

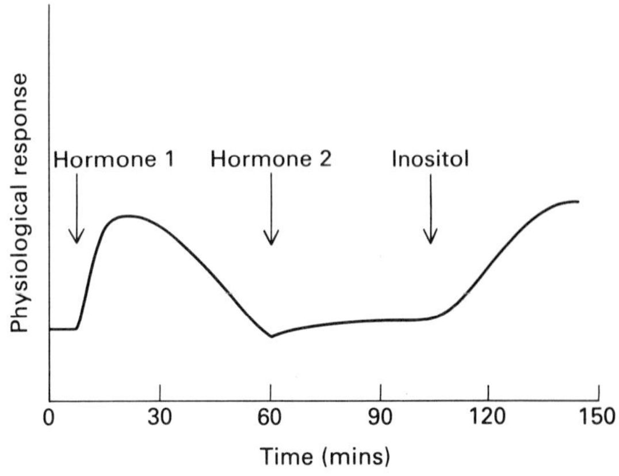

Fig. 9.5. Desensitization due to loss of membrane inositol phospholipid. The experiment is done using intact cells. Response to the first hormone declines over a period of about one hour and the response to other agonists is also lost. Addition of inositol leads to a gradual restoration of the response.

ponse persists and also affects all other agonists which stimulate phosphatidylinositol turnover in that cell type. The response can be restored by incubating the cells in the presence of inositol. This suggests that the depletion of the phosphatidylinositol pool causes the desensitization which is reversed when inositol for phosphatidylinositol synthesis is applied. From this, the clear conclusion is that the breakdown of phosphatidylinositol is necessary for the physiological response.

A second line of evidence comes from the relationship of phosphatidylinositol breakdown to receptor occupancy. In liver cells, the hormone concentration dependence of the stimulation of polyphosphoinositide breakdown closely follows the hormone binding curve. Physiological responses, such as the activation of phosphorylase, require a much smaller extent of receptor occupany (Fig. 9.6). This implies that the increase in phosphatidylinositol breakdown is an early response closely related to the binding of the hormone (see Chapter 2).

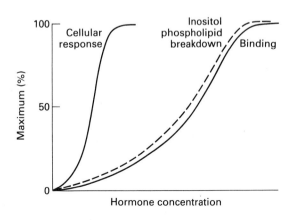

Fig. 9.6. Hormone binding in relation to physiological response and phosphatidylinositol breakdown. Most cellular responses are maximally stimulated at a low receptor occupancy corresponding to a low hormone concentration. Here, a receptor occupancy of about 5% gives maximum cellular response. In contrast, phosphatidylinositol breakdown corresponds quite closely to the extent of receptor occupancy.

9G Relationship to calcium mobilization

Agonists which increase phosphatidylinositol turnover also increase the cytoplasmic free Ca^{2+} concentration. This raises three possibilities.
1 The rise in Ca^{2+} causes the increase in phosphatidylinositol breakdown.
2 The increase in phosphatidylinositol breakdown causes the increase in cytoplasmic free Ca^{2+}
3 The two processes are parallel but not causally related.

At first sight, the suggestion that the increase in phosphatidylinositol breakdown is secondary to a rise in cytoplasmic Ca^{2+} is attractive since all the phosphodiesterases concerned are Ca^{2+} dependent. However, the concentrations of Ca^{2+} required to stimulate the enzymes in a cell extract are between 10 and 100 µmol/l. It is unlikely that Ca^{2+}-mobilizing agonists increase the cytoplasmic Ca^{2+} to concentrations much above 1 µmol/l.

More conclusive experiments have been done using mast cells, which respond to antigens and the lectin concanavalin A with increased phosphatidylinositol turnover. The physiological response is histamine release. In this case it has been possible to find conditions when phosphatidylinositol turnover is stimulated but there is no rise in the cytoplasmic free Ca^{2+} concentration. It seems likely, therefore, that the rise in phosphatidylinositol breakdown is not secondary to a rise in cytoplasmic Ca^{2+}. This leaves us with the possibilities that the rise in Ca^{2+} is secondary to a rise in phosphatidylinositol breakdown, or that the two processes are parallel but independent. In either case we must postulate a messenger molecule which is generated as a result of increased inositol phospholipid breakdown and then produces the cellular response.

9H Possible messenger molecules

9H1 Calcium

The first possible signal to consider is Ca^{2+} itself. This implies that the rise in cytoplasmic free Ca^{2+} is caused in some way by the increase in phosphatidylinositol turnover. It has been suggested that increased phosphatidylinositol breakdown allows Ca^{2+} to enter the cell across the plasma membrane, but this does not fit the observation that in many cell types hormones increase cytoplasmic free Ca^{2+} in the absence of Ca^{2+} in the external medium. A more attractive possibility is that phosphatidylinositol breakdown leads to the release of Ca^{2+} bound to the plasma membrane. Phosphatidylinositol 4,5-bisphosphate binds Ca^{2+} very tightly indeed and could provide a membrane-bound pool of Ca^{2+} for release. However, this would require a mechanism for the exclusion of Mg^{2+} which also binds very tightly and is present in the cytoplasm at a much higher concentration than Ca^{2+}. There is no direct evidence for such a mechanism, but membranes which have a high content of phosphatidyl 4,5-bisphosphate also have a high capacity to bind Ca^{2+}. Mobilization of Ca^{2+} from an intracellular organelle, as a consequence of phosphatidylinositol turnover, would require an intervening messenger.

9H2 Prostaglandins

The phosphatidylinositol pool in membranes has been shown to be rich in arachidonic acid (see Fig. 9.7). Arachidonate is the precursor for prostaglandins and leukotrienes, and it has been suggested that increased availability of the precursor leads to increased synthesis of prostaglandins which might act as messenger molecules. For this to be true, the supply of arachidonate must be limiting for prostaglandin biosynthesis. Prostaglandins typically act upon other cells close to the cell producing them, although the possibility that there may be affects on the producing cell has often been suggested.

Fig. 9.7. Structure of arachidonate.

$$CH_3(CH_2)_4CH= CHCH_2CH= CH-CH_2CH= CHCH_2CH= CH(CH_2)_3COO^-$$

9H3 Diacylglycerol

Recently, a protein kinase has been purified from platelets which is activated by diacylglycerol. It has been given the name C-kinase. In platelets, the kinase appears to phosphorylate a protein with a molecular weight of 40 000. The phosphorylation of this protein follows the release of diacylglycerol from phospholipids caused by stimulation of the cell by thrombin. It is assumed that the diacylglycerol is derived from the breakdown of phosphatidylinositol. The purified kinase can be stimulated by added diacylglycerol, and also requires Ca^{2+}. These results are very interesting, but it is difficult to assess their significance. The function of the 40 000 molecular weight protein which is phosphorylated is not known, and it is not known if this is a widespread mechanism occurring in many different cell types. C-kinase is also activated by phorbol esters which promote growth and malignant transformation of cells (see Chapter 12).

9H4 Inositol phosphates

Inositol phosphates are obvious candidates for the second messenger since they are the direct product of phosphatidylinositol breakdown. Inositol 1,4,5-trisphosphate has received the most attention as it is thought to be the earliest product. During the last two or three years, a lot of evidence has accumulated, supporting a role for inositol trisphosphate in controlling cytoplasmic Ca^{2+} concentration. The source of the initial rise in cytoplasmic Ca^{2+} in response to hormones is thought to be the endoplasmic reticulum. The addition of inositol trisphosphate to endoplasmic reticulum preparations from both liver cells and pancreatic acinar cells causes the release of Ca^{2+}. Inositol trisphosphate carries a very large negative charge and will not therefore pass across the plasma membrane. This problem has been overcome by permeabilizing cells with plasma membrane-specific detergents (see Chapter 6). If this is done, the addition of inositol trisphosphate causes release of Ca^{2+} from an intracellular source, which can be identified with a fair degree of confidence as the endoplasmic reticulum.

9J Conclusions

It seems very likely that the increase in phosphatidylinositol breakdown is important in the mechanism of action of calcium-mobilizing agonists. However, it is also clear that at present we are nowhere near a detailed understanding of the mechanisms involved, and the situation is rather

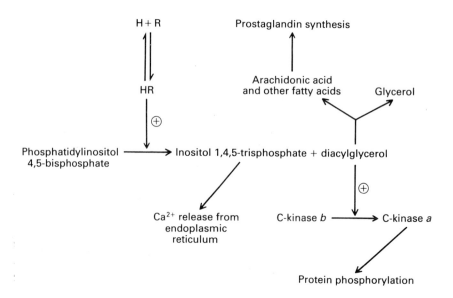

Fig. 9.8. Possible mechanisms for involvement of inositol 4,5-bisphosphate breakdown in hormone action.

confused. Figure 9.8 attempts to summarize the proposals which have been put forward. There is strong evidence for inositol trisphosphate acting as a second messenger for hormones by promoting the release of Ca^{2+} from the endoplasmic reticulum. The circumstantial evidence for a role for the diacylglycerol-activated protein kinase C is strong and will be considered further in Chapter 12. However, the details of the role of protein kinase C are very poorly understood. In the case of acute hormone effects, such as the response to α-adrenergic receptors, a reasonably clear picture of the role of phosphatidylinositol breakdown and its connection to Ca^{2+} mobilization is beginning to emerge. Where phosphatidylinositol breakdown and Ca^{2+} mobilization are associated with the control of long-term effects, such as cell growth and proliferation, the importance of the many possible mechanisms is much less obvious. This is discussed in Chapter 12.

Suggestions for further reading

Berridge M.J. (1982) A novel cell signalling system based on the interactions of phospholipid and calcium metabolism. In *Calcium and Cell Function* (Ed. by W.Y. Cheung), vol. 3, pp. 1–37. Academic Press.

Berridge M.J. (1984) Inositol triphosphate and diacylglycerol as second messengers. *Biochem J*, **220**, 345–360.

Williamson J.R., Cooper R.H., Joseph S.K. & Thomas A.P. (1985) Inositol trisphosphate and diacylglycerol as intracellular second messengers in liver. *Am J Physiol*, **17**, C203–C216.

Cell Calcium (1982) vol. 3 contains several articles covering most aspects of phosphatidylinositol metabolism.

Chapter 10 Mechanisms of action of insulin

10A Introduction

Insulin was first discovered by Banting and Best in the 1920s. By the 1930s several physiological effects of insulin had been characterized, notably its ability to accelerate the transport of glucose into many types of cell. A fairly accurate picture of the physiological effects of insulin was achieved by the early 1950s, but the elucidation of its mechanism of action has proved to be very difficult. It was appreciated quite early on that the action of insulin depended upon its binding to a cell-surface receptor. However, it is only during the last ten years or so that any sort of picture has emerged of the events which link the binding to subsequent effects on cell metabolism. Several mechanisms have been proposed, and it is still not clear which of these are important.

One of the main difficulties of studies on the mechanism of action of insulin is that, until very recently, no response following insulin binding had been demonstrated in a broken-cell preparation. The same problem arises with the Ca^{2+}-mobilizing hormones (see Chapter 6), and this contrasts with the situation where a hormone activates adenylate cyclase which can be readily measured in an isolated plasma membrane preparation. A further difficulty in elucidating the mechanism of action of insulin arises from the very wide range of different metabolic responses to the hormone.

10B Effects of insulin on cell metabolism

Table 10.1 summarizes the known effects of insulin in mammalian cells. All the responses are essentially anabolic, and insulin is the major hormone promoting anabolic metabolism, particularly rapid responses. Apart from this, the nature of the effects varies very widely, both in terms of the insulin concentration required to produce a response, and with regard to the length of time needed for a response. At one extreme, the acceleration of glucose transport or the inhibition of lipolysis require an insulin concentration of between 10^{-11} and 10^{-10} mol/l, and are maximal in less than a minute. At the other extreme, the growth-promoting effects of insulin, such as the stimulation of DNA replication, require four hours to be detectable, and about twenty-four hours to reach a maximum. The concentration of insulin needed for half-maximal stimulation is about 10^{-7} mol/l. There is a wide range of effects which fall between these two extremes. In most, but not all, cases, the longer term effects require higher concentrations of hormone.

If we make the assumption that there is only one type of insulin receptor, it follows that the first event resulting from insulin binding will

Table 10.1. Effects of insulin

Effect	Insulin K_a (mol/l)	Time for onset
Stimulation of glucose transport	5×10^{-11}	1 min
Stimulation of amino acid transport	10^{-9}	5 min
Changes in enzyme phosphorylation (glycogen synthetase, PDH, acetyl CoA carboxylase)	2×10^{-10}	3 min
Inhibition of lipolysis	10^{-11}	1 min
Activation of protein synthesis	10^{-9}	5 min
Activation of RNA synthesis	5×10^{-9}	10 min
Activation of DNA synthesis	10^{-7}	12 h

be the same whatever the final metabolic response. However, the very wide variation in both time course and dependence on insulin concentration, suggests the initial binding response may give rise to a number of different mechanisms which then result in the different final effects. If this is the case, it would explain why several different mechanisms have been proposed for the action of insulin, since individual workers tend to concentrate on a single cell type. A particular type of insulin response may be prominent in one cell type while in other cells different mechanisms are important.

10C Receptor binding

The first step in the mechanism of action of insulin is the binding of the hormone to its cell-surface receptor, and it is generally accepted that insulin does not need to enter the cell to produce its effects. There are a large number of analogues of insulin available. Artificial analogues can be made by chemical modification of insulin. Natural analogues are also abundant in that insulin has been purified from a large number of species, and the hormone from a foreign species will have a different potency from the insulin native to the species under study. Insulin from foreign species may be either more or less effective than the normal insulin. There is also a whole family of insulin-like growth factors which too can function as natural insulin analogues. There is a very good correlation between the ability of insulin analogues to bind specifically to a receptor and their ability to produce a physiological response. This is strong evidence in support of a key role for the initial binding interaction.

One of the earliest approaches used to study the mechanism of action of insulin was to attempt to relate insulin binding to the physiological response. Many of the first studies in the late 1960s were done on the isolated fat cell. This preparation has many advantages: it is homogeneous containing a single cell type, easily separated from the medium by flotation, and there are several well-defined and easily measured effects of insulin on fat-cell metabolism. The binding studies led to two different views of the nature of insulin binding. One group claimed to identify two classes of binding site for insulin, of high and low affinity. They claimed that the binding affinity of the high-affinity sites corresponded exactly to the concentration of insulin needed to stimulate glucose transport. To put this another way, they claimed that the same

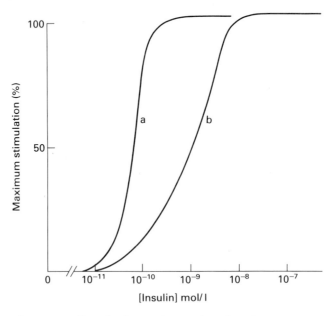

Fig. 10.1. Effect of reducing the number of insulin receptors on the activation of glucose metabolism by insulin: (a) shows the concentration dependence of the stimulation by insulin of glucose transport in control fat cells; (b) shows the effect of removing 90% of the insulin receptors by treatment with trypsin. The maximum response is not affected but the concentration of insulin required increases by a factor of 10.

insulin concentration (5×10^{-11} mol/l) gave 50% receptor occupancy for the high-affinity receptor, and 50% of maximum stimulation of glucose transport. Other groups suggested that there was a single class of insulin receptors with a much lower affinity, and that glucose transport was half-maximally stimulated at about 2% receptor occupancy.

As discussed in Chapter 2, there is no reason to expect that a response will follow the binding curve unless it is very early in the sequence of events after binding. The number of insulin receptors on the cell surface can be reduced by a brief exposure to the proteolytic enzyme trypsin. If the stimulation of glucose transport corresponds exactly to the receptor occupancy, the maximum possible stimulation will be reduced as the receptor number is reduced. If, on the other hand, maximal stimulation of transport requires a low receptor occupancy, maximal stimulation will not be affected until a large proportion of the receptors are destroyed. However, the concentration of insulin needed to produce an effect will be increased, since a larger proportion of the remaining receptors will need to be occupied to obtain the same total number of occupied receptors (Fig. 10.1). It was found that 90% of the receptors could be destroyed without affecting the maximal stimulation of glucose transport by insulin, but the concentration of insulin needed for the maximal response was increased by a factor of 10. This is convincing evidence in support of the low occupancy hypothesis.

10D Negative cooperativity in binding

Insulin binding does not follow simple saturation kinetics. The usual way to analyse hormone binding is by the Scatchard plot. A simple homogeneous class of binding sites gives a linear plot. Insulin binding gives a curved plot (Fig. 10.2) which can be interpreted in two ways: either there are two or more classes of binding site with different affinities, or there is a single class of binding sites that display negative cooperativity. Negative cooperativity implies that there is an interaction within a homogeneous population of binding sites, so that, as occupancy increases, binding to the unoccupied sites becomes progressively more difficult.

The evidence is in favour of negative cooperativity. At low temperatures, the dissociation of bound insulin follows first-order kinetics, indicative of a single class of binding site. Some insulin analogues show negative cooperativity of binding in a similar way to native insulin, but others are able to produce an insulin response but show no negative cooperativity. However, both types of analogue appear to compete for the same receptors. Negative cooperativity of insulin binding is quite

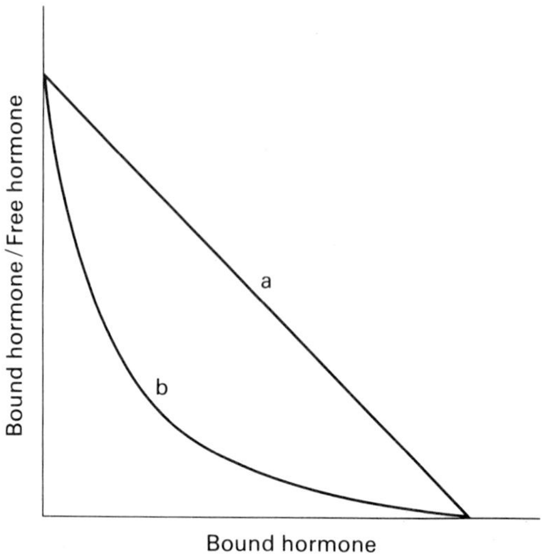

Fig. 10.2. Scatchard analysis of binding. (a) With a single homogeneous population of binding sites. (b) With either two or more types of site with different affinity, or a single homogeneous population which displays negative cooperativity of binding.

well established but it is not clear if it has any importance in the physiological response, since rapid effects of insulin require hormone concentrations well below those needed to detect negative cooperativity.

10E The insulin receptor

10E1 Structure of the receptor

The insulin receptor has been purified from a number of different tissues. Methods vary, but all involve the solubilization of plasma membranes using a non-ionic detergent such as Triton X100. A protein can be isolated which binds insulin with about the same affinity as the native receptor in the plasma membrane, and still reacts with antibodies to the insulin receptor. Determinations of molecular weight in detergent-dispersed preparations are a little uncertain since they involve assumptions about the amount of bound detergent. Most reports agree, however, that the molecular weight is about 350 000.

The subunit composition of the receptor has been analysed by SDS gel electrophoresis. In partially purified preparations, the detection of the receptor component was a problem since, as would be expected, SDS destroys the insulin-binding capacity. This was overcome by cross-linking radioactive insulin to the receptors with chemical cross-linking agents before solubilization. From this it emerged that the receptor contains two types of subunit. One has a molecular weight of 130 000 and is readily labelled by insulin in the presence of cross-linking agents.

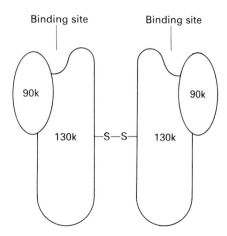

Fig. 10.3. Model of insulin receptor structure. This is based on the subunit composition revealed by SDS gel electrophoresis of radiolabelled insulin, chemically cross-linked to the receptor.

The other has a molecular weight of 90 000 and is less readily labelled. If disulphide bridges are not broken before running the gel, a single species with a molecular weight of 350 000 appears. This has led to the suggestion that the receptor consists of two heavy (130 000) and two light (90 000) chains linked by disulphide bridges (Fig. 10.3). A problem with this is that the total molecular weight of 350 000 is too small to account for the subunit molecular weights which would imply a total molecular weight of 440 000. This has led to the suggestion that the small species has a molecular weight of 45 000. Alternatively, the receptor may run anomalously on SDS gel electrophoresis. Large hydrophobic glycoproteins, such as the insulin receptor, frequently give inaccurate molecular weight determinations by this method.

The disulphide bridges appear to be important in the function of the receptor. Low concentrations of sulphydryl oxidants, notably peroxide, mimic the action of insulin, as do certain other sulphydryl agents such as cysteine. Reagents such as N-ethyl maleimide block the action of insulin, but this is less convincing since NEM inhibits many different cellular processes.

10E2 Glycoprotein nature of the receptor

Two lines of evidence suggest that the insulin receptor is a glycoprotein with the sugar component at the cell surface. At low concentrations, enzymes which hydrolyse polysaccharides, such as neuraminidase, mimic the action of insulin in fat cells. At higher concentrations, neuraminidase causes a very marked stimulation of glucose transport, and the effect of insulin on fat-cell metabolism is lost. The binding of insulin is not affected by neuraminidase, but if β-galactosidase is added as well as neuraminidase, both the cellular response and hormone binding are lost.

The second line of evidence involves the use of plant lectins. These are proteins which bind to sugar residues on the cell surface. Lectins which bind to N-acetyl glucosamine, mannose or galactose, also bind to detergent-solubilized preparations of the insulin receptor, and exert an insulin-like effect on metabolism. This has been extensively studied with the lectin concanavalin A (con A) which binds to mannose residues. Con A mimics many of the effects of insulin, but does not inhibit insulin binding. It does, however, have the interesting effect of preventing negative cooperativity in insulin binding.

10E3 The role of receptor cross-linking

Lectins tend to cross-link sugar-containing proteins on the cell surface, so the observation that lectins can mimic the action of insulin led to the idea that cross-linking of insulin receptors to each other or to other proteins might be important in the action of insulin. Con A is also able to stimulate DNA replication in lymphocytes and it is thought that cross-linking of surface proteins may be important in this response as well.

Con A is tetravalent, containing four binding sites for sugar. Monovalent fragments can be generated which still bind to the sugar but have lost the potential to cross-link. Such fragments are not insulin mimetic. Con A would be expected to cross-link many membrane proteins, not just specific receptors. Taken together with the observation that con A does not affect the binding of insulin to its receptor, this might suggest that the cross-linking effect is rather general and not specific to the insulin receptor. Antibodies to the insulin receptor were used to examine the effects of cross-linking only the insulin receptors.

Insulin-receptor antibodies were obtained from the serum of patients with a rare form of diabetes. The diabetes results from the presence in the blood of antibodies to the patient's own insulin receptors. The antibodies were effective against the insulin receptors of all human cells and also most other mammalian cells. Insulin prevented the binding of the antibody which bound to the protein identified as the insulin receptor in detergent-dispersed preparations. This shows that the effect of the antibody results from binding to the receptor, rather than from a secondary effect resulting from binding to another membrane protein.

Insulin-receptor antibody mimics a large number of insulin responses, the major exception being the stimulation of DNA biosynthesis. The most extensively studied system is the isolated fat cell where the insulin-receptor antibody is able to produce the full range of insulin effects. The

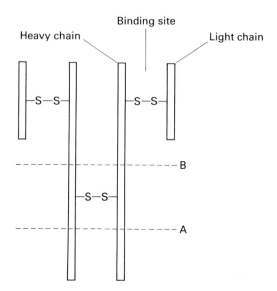

Fig. 10.4. Sites of proteolysis to form different types of anitbody fragment. Proteolytic cleavage at A leaves the link between the heavy chains intact and yields a divalent f(ab)$_2$ fragment. Proteolytic cleavage at B causes separation of the heavy chains yielding two monovalent fab' fragments.

responses measured are mostly rapid, acute effects of insulin; however, many are intracellular rather than plasma membrane events, and many are independent of the presence of glucose in the medium and therefore of the stimulation of glucose transport by insulin. The ability of the antibody to mimic the acute effects of insulin gives further support to the view that insulin exerts its effects through a single class of receptors which are immunologically identical.

The preparation of fragments of antibodies by limited proteolyis is a standard technique. It can be used to prepare fragments which are either divalent or monovalent, providing a means of examining the role of cross-linking in the effect of the antibody. The native antibody is divalent, and can therefore act as a cross-linker. Depending upon the site of proteolytic cleavage, f(ab)$_2$ fragments which are still divalent, or fab' fragments which are monovalent, can be formed (Fig. 10.4). The divalent f(ab)$_2$ fragments retain the ability to mimic the action of insulin like the intact antibody. The monovalent fab' fragments which clearly cannot cross-link the receptor are also unable to mimic the effects of insulin. It is clear that they still bind to the insulin receptors since they both compete with insulin for binding and act as competitive antagonists of the action of insulin. It seems, therefore, that in the case of the antibody, binding to the receptor is not sufficient to produce a response, but cross-linking is necessary. Antibodies to the monovalent fab' fragments will cross-link the bound fab' fragments and thereby restore the ability of the fab' fragments to cross-link the receptors (Fig. 10.5). This also restores the insulin mimetic effect.

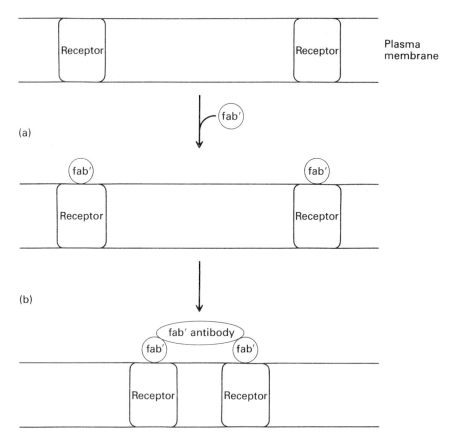

Fig. 10.5. Restoration of cross-linking with fab' fragments on addition of antibodies to fab'. (a) Addition of fab' leads to binding to the receptor, but it does not cross-link since it is monovalent. (b) Addition of divalent antibody to fab' restores the cross-linking.

It is clear that the effects of antibodies to the insulin receptor require the cross-linking of the receptors. It does not necessarily follow from this that the effects of insulin itself require receptor cross-linking. This is a difficult question to approach experimentally, but there is some evidence that cross-linking of the receptor is important in the action of the hormone. The dose–response curve to insulin can be shifted to the left by the addition of antibodies to insulin (Fig. 10.6). The experiment is analogous to the experiment with fab' fragments shown in Fig. 10.5. Antibodies to insulin might be expected to promote the cross-linking of insulin bound to receptors. Since this increases the effectiveness of insulin, reducing the concentration of the hormone required for a given response, it suggests that receptor cross-linking may be important in the action of insulin itself. Fluorescent labelling of insulin and of insulin-receptor antibodies has been used to follow aggregation of the insulin receptor on the surface of lymphocytes. Both the hormone and the antibody form patches and caps on the cell surface. This is usually taken to be a reflection of protein cross-linking, and is good evidence that

153 Mechanisms of action of insulin

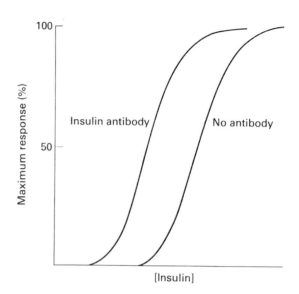

Fig. 10.6. Effect of insulin antibodies on the response to insulin. Addition of insulin antibody reduces the concentration of insulin needed for the response, implying that cross-linking is important in the mechanism of action of insulin.

insulin can cross-link its receptors. However, it should be pointed out that patching and capping are slow processes and require much higher concentrations of insulin than are needed for rapid hormone effects. It is unlikely that the formation of such large aggregates of the receptor is necessary for the action of insulin.

Insulin itself shows little tendency even to form dimers at physiological concentrations, and it is not at all clear how the hormone could cross-link receptors. A conformational change in the receptor, consequent upon insulin binding, could promote cross-linking. The effects of sulphydryl reagents (see Section 10E1) suggest that the breaking and reforming of disulphide bridges might be involved.

There is no evidence that insulin actually promotes the aggregation of receptors on the cell surface. If such an association was a necessary part of the mechanism of action, drugs such as colchicine, cytochalasin or vinblastine which disrupt the function of the cytoskeleton should interfere with the action of insulin, but they have little effect. The location of insulin receptors on the cell surface can be examined using insulin with ferritin attached as an electron-dense stain for electron microscopy. Ferritin insulin has full biological activity and must therefore bind specifically to insulin receptors. Using this method, it appeared that most insulin receptors occur in clusters of from two to six although there are also a few single receptors. Pretreatment with insulin did not affect the relative proportions of clustered and single receptors. It seems likely that cross-linking is an essential step in the mechanism of action of insulin, and that insulin promotes cross-linking between receptors which are already associated in clusters in the membrane.

10E4 Self-phosphorylation of the insulin receptor

Recent studies have shown that insulin binding to the insulin receptor activates a protein kinase. The kinase activity appears to be a function of the receptor itself, and catalyses the incorporation of phosphate into the small, 90 000 molecular weight chain of the receptor. The phosphorylation occurs on a tyrosine residue rather than upon the more usual serine or threonine (Fig. 10.7). The receptor is also able to phosphorylate tyrosine residues on a number of other proteins. These observations are very interesting indeed. Tyrosine phosphorylation is unusual, the other known tyrosine kinases being growth-factor receptors or the protein products of oncogenes. As yet, the significance of the insulin-receptor effect is unknown.

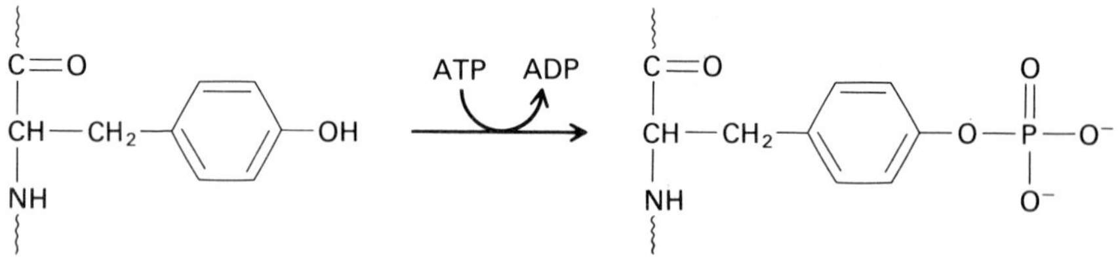

Fig. 10.7. The tyrosine kinase reaction.

10E5 Receptor-mediated growth effects

The growth-promoting effects of insulin take place over longer periods of time and require much higher concentrations of hormone than the acute metabolic effects. The greater length of time needed may simply reflect the fact that it takes longer to detect a change in the rate of protein synthesis than, for example, in the rate of glucose transport. The concentration of insulin required for some of the growth effects can be as much as 10^{-7} mol/l—three orders of magnitude higher than the concentration needed for acute metabolic responses. This raises two questions. First, are these growth-promoting effects physiologically relevant? It seems unlikely that they are, since the insulin concentrations needed are far in excess of those ever observed in the circulation. Second, does insulin produce its growth-promoting effects through the same receptors as mediate the acute effects? This also seems unlikely since binding studies suggest that the insulin receptor should be saturated at concentrations of insulin below those needed for half-maximal stimulation of the growth responses. This led to the suggestion that the growth effects of insulin were mediated by insulin-like growth factor (IGF) receptors while, conversely, the acute effects of IGF were mediated by insulin receptors. Monovalent fab' fragments of insulin-receptor antibodies can be used as specific competitive inhibitors of insulin binding to its receptor. They were found to inhibit the acute effects, but not the growth-promoting effects of both insulin and IGF. This supports the idea that the growth-promoting effects of insulin *in vitro* are mediated through a separate population of receptors which normally respond to IGF. However, it should be emphasized that physiological levels of insulin are required before the effects of growth factors can be expressed. Insulin *in vivo* probably has a permissive rather than a direct role to play in the stimulation of growth.

A good deal is known about the insulin receptor. The binding characteristics of insulin have been well worked out. The structure of the receptor is partly elucidated and we have some idea of the chemical events which follow directly upon the binding of the hormone. However, it is still unclear how these events relate to the metabolic response.

10F The insulin second messenger

10F1 The need for a second messenger

One of the most obvious effects of insulin is its ability to increase the rate of glucose transport into most cells, the major exception being the liver. In fat cells or skeletal muscle, the rate of influx of glucose can be increased by as much as 10- to 20-fold. Obviously this in itself has a marked effect on the metabolism of the cell, and it was suggested that most of the other effects of insulin were secondary to the increase in glucose transport. If insulin was only affecting membrane transport of glucose, it was possible for this to be a direct effect of the receptor and there was no obvious need for a second messenger to be released into the cytoplasm. In a way, glucose itself was suggested to be the second messenger. However, in the liver, glucose transport is not affected by insulin which, nevertheless, has profound effects on liver cell metabolism. Furthermore, in cell types where insulin does accelerate glucose transport, many of the other effects of insulin can be observed in the absence of extracellular glucose. Many of the effects of insulin involve changes in the activity of cytoplasmic, or, in at least one case, intramitochondrial enzymes, indicating that there must be at least one second messenger for insulin. As yet, there is no clear evidence of the identity of the second messenger, so the following sections can only attempt to review the various proposals which have been made. At the risk of

being repetitive, it might be useful to restate the criteria for the identification of a second messenger.
1 The hormone must change the concentration of the second messenger.
2 The second messenger must alter cellular processes in a direction consistent with the action of the hormone.
3 There must be a mechanism for the removal of the messenger.
4 The appropriate sequence of events should be established.

These criteria need to be borne in mind when considering candidates for the insulin second messenger.

There have been two types of hypothesis for the identify of the insulin second messenger. Insulin binding leads to internalization of the insulin receptor together with the bound insulin. One possibility is that either the insulin itself, or the receptor or component of the receptor, might be the second messenger. The second possibility is that the binding of the hormone to the receptor leads to the activation or inhibition of a membrane enzyme or transport process, or to the release of a factor bound to the inner surface of the plasma membrane.

10F2 Receptor-mediated internalization

It was found that radioactively labelled insulin becomes irreversibly associated with target cells over a period of time (Fig. 10.8). The location of the insulin was examined by light microscopy using fluorescent analogues of insulin, and by electron microscopy using ferritin insulin conjugates. It was found that the insulin first formed patches on the cell surface and was then internalized, presumably still bound to the recep-

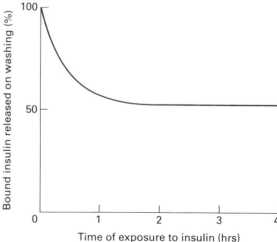

Fig. 10.8. Effect of time of exposure to hormone on the proportion of bound insulin that can be removed by washing. Insulin gradually becomes irreversibly associated with the cell. The proportion varies from 15 to 50% and the time taken from 30 minutes to 4 hours depending upon the cell type.

tor, by a process of endocytosis. The first criterion for a second messenger can be said to be established in that the binding of insulin has resulted in the appearance of the putative second messenger, insulin itself, or a component of the receptor in the cytoplasm.

Considering the second criterion for a second messenger. None of the metabolic effects of insulin can be produced by adding insulin to any type of broken-cell preparation. In the case of the receptor, or a possible component derived from the receptor, there is no evidence one way or the other. A putative second messenger of this type has not been isolated and cannot therefore be tested.

If we consider the third criterion; internalized insulin is rapidly broken down by cellular proteases. The cell certainly contains enough proteolytic activity to break down either insulin or its receptor by proteolysis, so a means exists for the removal of the proposed second messenger.

Finally, the correlation between the physiological response to insulin and the internalization process can be considered. Both the concentration dependency and the time course correlate very poorly. Detectable internalization requires both much longer periods of time and much higher insulin concentrations than the acute effects of insulin. Against this it might be possible to argue that this is a problem of sensitivity, and that at the levels of insulin needed for the physiological response, internalization does take place, but is below the limits of detection of the methods available. A more conclusive approach is to block one or other of the two processes. Drugs active on the cytoskeleton, such as colchicine or cytochalasin, block the internalization process, but do not block acute effects of insulin. There are a number of ways to block either the physiological response or internalization. The correlation between the two processes is very poor with the exception of the insulin stimulation of DNA synthesis. As discussed above (Section 10E5), this is probably controlled via a different receptor from that promoting the acute effects.

It seems far more likely that the role of internalization is in down regulation or desensitization, rather than in the primary response to insulin. The physiological role is probably the removal of receptors in the face of abnormally high concentrations of insulin, in order to prevent an excessive response. This type of phenomenon is common to many different hormones.

10G Receptor-mediated production of a second messenger

Several different candidates have been considered for the second messenger for insulin action. These have included the cyclic nucleotides and Ca^{2+} ions, which are well established as second messengers for other extracellular agonists, as well as specific insulin second messengers.

10G1 Cyclic nucleotides

In many cell types, insulin antagonizes the action of hormones which are known to exert their effects through a rise in the level of cyclic 3′,5′-AMP. This led to the idea that insulin might use the same second messenger but in the reverse sense; that is, that it might produce its effects through a reduction in the levels of cyclic AMP. The idea is obviously attractive in that it would provide a single unified mechanism for the action of two major classes of hormones.

In liver cells and in fat cells, insulin does reduce the level of cyclic AMP, but only if the cyclic AMP level has previously been increased by glucagon or adrenaline. There is no evidence that insulin reduces the resting level of cyclic AMP, and in both cell types many effects of insulin are apparent in the absence of other hormones. In liver, glycogen synthesis and fatty acid synthesis are increased. In fat cells, pyruvate dehydrogenase is activated by dephosphorylation, basal lipolysis is reduced and, again, fatty acid biosynthesis is increased.

There is, therefore, little evidence that a lowering of cyclic AMP levels can be responsible for the primary effects of insulin. This leaves the question of whether reductions in cyclic AMP can explain the effects of insulin in antagonizing the action of glucagon and adrenaline. In the

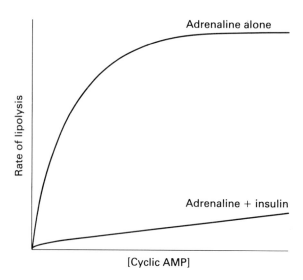

Fig. 10.9. Relationship of the rate of lipolysis in fat cells to the cyclic AMP concentration, in the presence and absence of insulin. Cyclic AMP levels are elevated by the addition of adrenaline in the absence or presence of insulin. In the presence of insulin the stimulation of lipolysis is markedly less.

fat cell, there is a very poor correlation between the effects of insulin in reducing adrenaline-stimulated lipolysis and in reducing cyclic AMP levels. If lipolytic rates are compared at the same level of cyclic AMP, the rate in the presence of insulin is always markedly lower (Fig. 10.9). In liver, on the other hand, there is a fairly good correlation between the ability of insulin to reduce cyclic AMP levels in the presence of glucagon, and its ability to inhibit glucagon-stimulated processes. A mechanism for the reduction of previously elevated cyclic AMP levels has also been identified. The addition of insulin, in the presence of ATP, to isolated liver plasma membranes results in the phosphorylation and activation of a membrane-bound cyclic AMP phosphodiesterase. The phosphorylation depends not only upon the presence of insulin but upon the presence of cyclic AMP. The levels of cyclic AMP required are comparable with those produced as a result of glucagon stimulation of liver cells.

Thus it appears that there is an insulin-activated, cyclic AMP-dependent protein kinase which phosphorylates and activates a cyclic AMP phosphodiesterase. This would account for the observation that cyclic AMP levels must be raised above basal for insulin to be able to exert an effect. This mechanism is of general interest and will be returned to later (see Section 10H6). It appears then that reductions in cyclic AMP levels may in some cases be important in the ability of insulin to antagonize the effects of hormones which activate adenylate cyclase. It does not provide an explanation for primary effects of insulin.

There is some evidence that insulin increases the concentration of cyclic GMP in fat cells and in liver. This led to the suggestion that cyclic GMP was the second messenger for insulin. This hypothesis also has a certain pleasing symmetry, with cyclic AMP as the second messenger for catabolic hormones, while cyclic GMP is the messenger for the main anabolic hormone. However, there is very little evidence that cyclic GMP can affect metabolic processes in a cell-free system in the appropriate direction. This could be the result of deficiencies in the cell-free preparation used. Another approach is to use other effectors, drugs or other hormones, which raise cyclic GMP levels in an intact-cell preparation and see if these produce an insulin-like response. It turns out that some do mimic insulin, some do nothing and others have effects opposite to insulin. This is still not conclusive since it is possible to argue that the effectors may also stimulate other mechanisms to produce the inappropriate effects. However, cyclic GMP does not stand up well to the basic criteria for a second messenger for insulin. While it has not been shown conclusively that it is not involved, the evidence in its favour is weak.

10G2 Ca^{2+} as a second messenger for insulin

There are many publications suggesting that Ca^{2+} is the second messenger for insulin. Most of the experiments relating to this proposal have been done using isolated fat cells or, in a few cases, skeletal muscle.

There is reasonable evidence that insulin does alter the cytoplasmic free Ca^{2+} level. In the fat cell, ^{45}Ca efflux from preloaded cells is stimulated. This is good evidence that the Ca^{2+} distribution in the cell is changed, but the source of the change and whether the cytoplasmic Ca^{2+} concentration is increasing or decreasing is difficult to determine (see Chapter 5). Unfortunately more direct methods, such as null point titration or the use of fluorescent indicators, are difficult to apply to fat cells because of their relative fragility and the presence of a fat globule taking up over ninety per cent of the cell volume.

Another approach has been to examine the effect of insulin on skeletal muscle contraction. Insulin does not induce contraction in a normal resting muscle. If, however, the muscle is induced to contract by exposure to a hyperosmolar medium, the addition of insulin induces further contraction, indicative of an increase in the cytoplasmic free Ca^{2+} level. This is convincing evidence for an increase in cytoplasmic free Ca^{2+} but it must be said that the conditions used to achieve it are rather artificial. In both cases there is no requirement for extracellular Ca^{2+} for the initial response, or indeed for the physiological effects of insulin. The Ca^{2+} would therefore have to be derived from an intracellular source. In liver cells, there is no evidence that insulin increases the cytoplasmic free Ca^{2+} concentration and this has been tested by direct methods. The evidence that insulin changes cytoplasmic Ca^{2+} is therefore equivocal; in the fat cell it probably does, in muscle it may, and in liver it probably does not.

The next question is, does an increase in Ca^{2+} in the cytoplasm produce an insulin-like effect? In isolated fat cells, a number of manipulations which would be expected to increase cytoplasmic free Ca^{2+} levels do mimic the metabolic response to insulin. Fat cells have a Na^+/Ca^{2+} exchanger in the plasma membrane so that a reduction in the Na^+ gradient by, for example, substituting external Na^+ for Li^+ should increase cytoplasmic Ca^{2+}. Making the medium hyperosmolar would be expected to have the same effect. Both these treatments mimic the action of insulin. Cooling the cells tends to increase the cytoplasmic Ca^{2+} as the pumps responsible for Ca^{2+} efflux are more temperature sensitive than the processes allowing influx. On restoration to normal temperature, the cells behave for a short time as if they had been stimulated by insulin. The addition of uncoupler to fat cells inhibits many of the insulin-sensitive metabolic pathways. However, in the presence of oligomycin to block the mitochondrial ATPase, the cytoplasmic ATP is maintained quite well by glycolysis. In part this is due to an acceleration of glucose transport, an insulin mimetic effect. Similar effects are seen in muscle. Uncouplers might be expected to cause the release of Ca^{2+} from the mitochondria.

There is a considerable body of evidence that a rise in cytoplasmic free Ca^{2+} mimics the effects of insulin in fat cells. However, the evidence remains circumstantial in that it has not been shown directly that the various manipulations do raise the cytoplasmic free Ca^{2+} concentration. In liver cells, agents which increase cytoplasmic free Ca^{2+} do not mimic the effects of insulin.

The next question to consider is whether or not enzyme activities are affected by Ca^{2+} in a direction appropriate to the physiological response. In fat cells there are a number of instances where this is the case. The mitochondrial enzyme, pyruvate dehydrogenase, is activated by insulin in fat cells. The activation results from dephosphorylation of the enzyme, and the phosphatase is activated by Ca^{2+}. As discussed in Chapter 5, a rise in cytoplasmic free Ca^{2+} might be expected to cause a rise in mitochondrial free Ca^{2+}. Also, pyruvate dehydrogenase activity in isolated mitochondria has been shown to increase when the extramitochondrial free Ca^{2+} is increased. This is certainly consistent with a role for an increase in cytoplasmic Ca^{2+} in the action of insulin on pyruvate dehydrogenase in fat cells, although other factors affect the phosphorylation state of the enzyme. It also provides evidence against a role for Ca^{2+} in the action of insulin on liver. Insulin does not increase PDH activity in liver cells, while vasopressin and α_1-adrenergic agonists which are known to increase liver cytoplasmic free Ca^{2+} do increase PDH activity.

A major argument against a role for Ca^{2+} as a second messenger for insulin arises from the control of glycogen metabolism. Insulin activates glycogen synthetase and inhibits glycogen breakdown. Phosphorylase kinase is activated by Ca^{2+} (see Chapter 14) which has exactly the opposite effect.

A number of mechanisms have been proposed by which insulin might change the cytoplasmic free Ca^{2+}. It has been suggested that it promotes the release of Ca^{2+} from binding sites on the inner surface of the plasma membrane and there is evidence that insulin decreases Ca^{2+} binding to fat-cell plasma membranes. This would allow a transient rise in cytoplasmic Ca^{2+}, but in the absence of a change in plasma membrane Ca^{2+} fluxes, the rise could not persist. There is some evidence that the Ca^{2+}-ATPase in fat-cell plasma membrane is inhibited by insulin. This would tend to increase cytoplasmic free Ca^{2+} (see Chapter 5).

10G3 Regulatory factor

Recently a regulatory factor has been isolated from fat cells and from skeletal muscle which can mimic some of the effects of insulin in a cell-free system. The factor is prepared by incubating isolated plasma membranes with insulin, removing the membranes by centrifugation and taking the supernatant. It can also be generated by insulin-receptor antibodies, or by con A which is believed to mimic the actions of insulin through the insulin receptor.

If fat-cell mitochondria are exposed to the factor, PDH is activated by a decrease in its phosphorylation state. In skeletal-muscle homogenates, the factor inhibits cyclic AMP-dependent protein phosphorylation and inhibits the dephosphorylation of glycogen synthetase. The factor appears to be a small acidic peptide with a molecular weight of 2000−4000. The proposal is that insulin, acting via its receptor, causes its release from binding sites on the inner surface of the plasma membrane.

This work is very interesting but is still at a rather preliminary stage. As yet, the factor has not been fully purified, and there is some disagreement as to its chemical properties. There remains the possibility that there may be more than one type of factor. Relatively few insulin effects have been shown to be reproduced by the factor so far, and the effects observed are small in relation to the size of the effects seen with insulin in the intact cell.

10H Mechanisms for alteration of enzyme activity by insulin

10H1 Glucose transport

Insulin markedly activates glucose transport in most cell types, the notable exception being the liver. The effect of the hormone is to increase the V_{max} of transport without affecting the concentration of glucose which gives half-maximal velocity. This is unusual in mechanisms involving a direct activation by an effector binding to an enzyme and suggests that insulin increases the number of functioning glucose transport proteins. To test this requires an assay for the presence of the glucose transporter itself, rather than the rate of glucose transport.

Mechanisms of action of insulin

Two methods have been used to measure the amount of glucose transporter. The first makes use of an inhibitor of glucose transport, cytochalasin B. Inhibition is competitive with glucose so that glucose-suppressible cytochalasin B binding can be used as an assay for the number of glucose transporters present. The second method involves solubilization and partial purification of the transporter which is then incorporated into artificial lipid vesicles. Glucose transport activity in the vesicles is measured and this can also be used as an assay for the amount of glucose transporter present. Both methods reveal the presence of glucose transporter in both the plasma membrane and endoplasmic reticulum of fat cells. The addition of insulin causes a redistribution of the transporter away from the endoplasmic reticulum to the plasma membrane.

The activity of the glucose transporter was determined in the same cell preparation by measuring the rate of entry of the non-metabolizable analogue, 3-O-methyl glucose. The time course of appearance of the transporter in the plasma membrane correlated well with the increase in the rate of 3-O-methyl glucose transport (Fig. 10.10) as did the dependence on insulin concentration. The results from both methods suggest that insulin increases the number of glucose transport proteins in the plasma membrane. The total number of cytochalasin B binding sites did not change suggesting that the activation of glucose transport results only from the redistribution of the transporter within the cell. However, the total activity measured after incorporation into lipid vesicles is increased by insulin, suggesting that the hormone may activate latent glucose transporters as well as causing a movement of transporters from

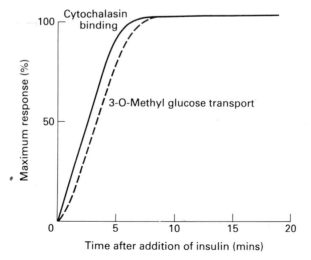

Fig. 10.10. Correlation of appearance of cytochalasin binding sites in the plasma membrane with activation of 3-O-methyl glucose transport by insulin.

the endoplasmic reticulum to the plasma membrane. Cytochalasin B binding may not distinguish between active and inactive transporters.

It appears that the movement of glucose transporters from the endoplasmic reticulum to the plasma membrane is an important factor in insulin stimulation of glucose transport and that an activation step may also be involved. This hypothesis would explain why insulin does not activate glucose transport to any extent in intact plasma membrane vesicles which have glucose transport activity. In the absence of endoplasmic reticulum, the source of additional transport protein is lost. It also explains why insulin affects only the V_{max} and not the K_m for transport of glucose and of many analogues of glucose.

The mechanism of this effect is not at all well understood. The current hypothesis suggests that vesicles containing glucose transporters cycle between the plasma membrane and the endoplasmic reticulum by fusion and pinocytosis. Insulin then either promotes the fusion or decreases the rate of pinocytosis, or both (Fig. 10.11). If true, this idea has

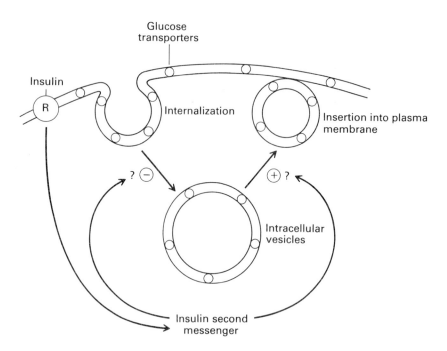

Fig. 10.11. Cycling of glucose transporters between the plasma membrane and endoplasmic reticulum.

important implications. As far as is known, insulin does not alter the concentration of other proteins in the plasma membrane. From this it follows that glucose transporters must be concentrated in specific areas of the plasma membrane which can be recognized for pinocytosis, and that glucose transporter-containing vesicles in the cytoplasm can be recognized for fusion. This effect could be mediated by the insulin receptor directly, or by a second messenger, but there is no evidence on this point.

The hypothesis must also accommodate the fact that many effectors other than insulin activate glucose transport in both fat cells and skeletal muscle. These include muscle contraction, α-adrenergic agonists, vasopressin and, provided that steps are taken to maintain cytoplasmic ATP levels, uncouplers of oxidative phosphorylation. It is notable that these agents and a number of others which produce the same effect are all thought to increase the cytoplasmic free Ca^{2+} concentration, and in the fat cell, at least, there is evidence that insulin changes the Ca^{2+} balance of the cell. However, Ca^{2+} is likely to be involved in a process involving the translocation of protein from one part of the cell to another. It is probably required for the response, but does not necessarily initiate it.

10H2 Enzyme phosphorylation

Hormones which function via a rise in cellular cyclic AMP levels clearly produce most of their effects by stimulating protein phosphorylation through the cyclic AMP-dependent protein kinase (see Chapter 4). The existence of calmodulin-sensitive protein kinases and of the diacylglycerol-sensitive C-kinase had led to the suggestion that the final metabolic effects of calcium-mobilizing hormones may also be mediated by protein phosphorylation (see Chapters 7 and 9). This has led to the idea that hormones in general, and perhaps other extracellular agonists, may alter metabolism as a consequence of changes in protein phosphorylation, and there has been a lot of interest in the possibility that this may also provide an explanation of the effects of insulin.

10H3 Whole-cell phosphorylation studies

There have been a number of studies in both fat cells and liver cells where the effect of insulin on protein phosphorylation in the intact cell has been measured. Cells are preincubated with $[^{32}P]$-P_i to label the ATP; hormones, including insulin, are added, and the total cell protein is separated by SDS electrophoresis. In both fat cells and liver cells the phosphorylation level of several proteins is changed by insulin; in some cases increased and in other cases decreased. This type of experiment lends support to the view that changes in protein phosphorylation state may be important in the action of insulin. Some of the enzymes whose phosphorylation state is changed have been identified.

10H4 Pyruvate dehydrogenase (PDH)

Pyruvate dehydrogenase was the first enzyme where the phosphorylation state was shown to be affected by insulin, the effect being confined to fat cells. Insulin leads to dephosphorylation of the pyruvate decarboxylase subunit of the enzyme leading to activation of the whole complex (see Chapter 16). There is good evidence that this is due to an activation of PDH phosphatase.

PDH is a rather special case since, in contrast to all the other enzymes affected, it is intramitochondrial, and the kinase and phosphatase are specific to PDH itself. As a result, while the effects of insulin on PDH are quite well characterized, this has not yielded much information about insulin action which is generally applicable.

10H5 Cytoplasmic enzymes

A number of cytoplasmic enzymes alter in phosphorylation state in response to insulin (see Table 10.2). In some cases the evidence for any physiological significance is poor or the effects are poorly characterized. However, the effects on glycogen synthetase and acetyl CoA carboxylase are well described.

Table 10.2. Soluble enzymes where the phosphorylation state changes in response to insulin

Enzyme	Change	Tissue
Glycogen synthetase	−	Liver, fat, muscle
Phosphorylase kinase	−	Liver
Phosphorylase	−	Fat, liver
Pyruvate kinase	−	Liver
Pyruvate dehydrogenase	−	Fat
Acetyl CoA carboxylase	+	Fat, liver
β-Hydroxymethylglutaryl-CoA lyase	−	Liver
Triacylglycerol lipase	−	Fat

Glycogen synthetase

The control of glycogen synthetase by phosphorylation is very complicated and is described in detail in Chapter 14. Insulin reduces phosphorylation at the sites which are phosphorylated by cyclic AMP-dependent protein kinase. This probably results from its ability to lower cyclic AMP levels in opposition to glucagon or adrenaline. However, insulin also specifically reduces phosphorylation at site 3. This site is the substrate for glycogen synthetase kinase 3 which is not affected by either cyclic AMP or Ca^{2+} ions. The reduction in phosphorylation could result from either the activation of protein phosphatase or the inhibition of glycogen synthetase kinase 3. Known protein phosphatases have a rather broad specificity with respect to the phosphorylated proteins which they act upon, while the kinase is specific for site 3 in glycogen synthetase, as is the effect of insulin. It seems likely, therefore, that the effect of insulin results from inhibition of kinase 3. It would be very interesting to know if glycogen synthetase kinase 3 has any other protein substrates.

Acetyl CoA carboxylase

In both adipose tissue and liver, acetyl CoA carboxylase activity is increased by insulin and decreased by a rise in cyclic AMP levels. The rise in cyclic AMP is accompanied by a substantial increase in the phosphorylation of the enzyme, presumably as a result of activation of the cyclic AMP-dependent protein kinase. Insulin causes a small but consistent increase in the total phosphorylation of the enzyme amounting to about ten per cent. Insulin caused a fivefold increase in phosphorylation at a single site which was distinct from the site phosphorylated in response to a rise in cyclic AMP.

Both these enzymes illustrate the need to analyse hormone-dependent protein phosphorylation at the level of a particular site. Both enzymes provide a potential assay for insulin-regulated protein kinases. In the case of acetyl CoA carboxylase, the kinase appears to be a plasma membrane enzyme which is independent of either cyclic AMP or Ca^{2+} ions, but as yet there is no direct evidence that insulin activates this enzyme.

10H6 Plasma membrane phosphorylation: cyclic AMP-dependent phosphodiesterase

Insulin has been shown to change the level of phosphorylation of a number of proteins in the plasma membrane in both liver cells and fat cells. In liver cells the effect can be observed using isolated plasma membranes and ATP. Insulin decreases the phosphorylation of two integral proteins of molecular weight 140 000 and 80 000 and increases the phosphorylation of three peripheral proteins of molecular weights 52 000, 28 000 and 14 000. All five phosphorylations are cyclic AMP dependent. The 52 000 molecular weight peripheral protein has been identified as a low K_m cyclic AMP phosphodiesterase which is activated by phosphorylation. Thus there appear to be at least two plasma membrane-bound, cyclic AMP-dependent protein kinases. One is apparently activated by insulin, while the other is inhibited. The effect on cyclic AMP phophodiesterase phosphorylation and activity suggests a mechanism by which insulin might antagonize the effect of glucagon by lowering cyclic AMP levels, since the activation of the phosphodiesterase is dependent upon elevation of the cyclic AMP concentration. The identity and function of the other four proteins whose phosphorylation state changes is unknown.

10J Protein kinases involved in insulin action

It is clear that the effects of insulin on protein phosphorylation are complex and probably involve several different protein kinases. In various different tissues insulin has been shown to affect two membrane-bound cyclic AMP-dependent protein kinases, soluble cyclic AMP-dependent protein kinase, by reducing the concentration of cyclic AMP, and at least two different protein kinases, one membrane-bound and one soluble, which are not affected by either cyclic AMP or Ca^{2+} ions. There is also the possibility that insulin might affect protein phosphatase activity and in the special case of PDH phosphatase this is well established. The mechanisms by which insulin produces these effects are largely unknown.

Finally, the role of the tyrosine-specific protein kinase activity of the insulin receptor itself should be considered. As well as phosphorylating itself, the occupied receptor has been shown to phosphorylate tyrosine residues on a number of other proteins. The physiological significance of this is unknown, and it should be emphasized that where insulin has been shown to alter the activity of an enzyme by changing its phosphorylation state, the phosphorylation always takes place on a serine residue. However, tyrosine phosphorylation is an unusual reaction and there are few known tyrosine kinases. It seems likely that they are important. In the case of the insulin receptor, the tyrosine kinase activity could be responsible for the initial effects of insulin perhaps by altering other protein kinase activities by phosphorylation. Alternatively, self-phosphorylation of the receptor could be a stage in the activation process. However, there is, as yet, no evidence to determine the role of the receptor tyrosine kinase activity.

10K Conclusions

A few years ago a famous cartoon was published showing a lecturer standing in front of a picture of a large black box saying 'Now we really understand how insulin works, it binds to the cell surface, then something happens then we get a response'. The present situation is somewhat better since a lot of detailed information about the action of insulin is now available. However, as yet, no coherent picture of the mechanism of action has emerged and a cynic could say that our confusion is merely at a more sophisticated level. The problem probably arises as a result of the great diversity of action of insulin. There may be a single initial mechanism resulting from the binding of insulin to its receptor. On the other hand there is no reason why there should not be several changes in the receptor, each promoting a different set of effects. Certainly, at the level of the final metabolic response, it is clear that several different effectors of enzyme activity are involved.

Suggestions for further reading

Denton R.M., Brownsey R.W. & Belsham G.J. (1981) A partial view of the mechanism of action of insulin. *Diabetologia*, **21**, 347−362.

Gliemann J. & Rees W.D. (1983) The insulin sensitive hexose transport system in adipocytes. In *Current Topics in Membranes and Transport* (Ed. by A. Kleinzeller and B.R. Martin), vol. 18, pp. 359−381. Academic Press.

Goldfine I.D. (1981) Effects of insulin on intracellular functions. In *Biochemical Actions of Hormones* (Ed. by G. Litwack), vol. 8, pp. 274−307. Academic Press.

Houslay M.D. & Heyworth C.M. (1983) Insulin in search of a mechanism. In *Trends in Biochemical Sciences*, vol. 8, pp. 449−452.

Kahn C.R., Baird K.L., Flier J.S., Granfield C., Harmon J.T., Harrison L.C., Karlson F.A., Kasuga M., King G.L., Lany U.C., Podskalny J.M. & Van Obberghen E. Insulin receptors, receptor antibodies and the mechanism of insulin action. *Recent Progress In Hormone Research*, **37**, 477−534.

Chapter 11 Mechanism of action of steroid hormones

11A Introduction

Steroid hormones are involved in the regulation of long-term processes such as development, sexual maturation and pregnancy. The glucocorticoids are involved in the response to starvation and the mineralocorticoids in the maintenance of water and salt balance.

It was established many years ago that the major effects of steroid hormones follow from their ability to induce or repress the synthesis of specific enzymes. The glucocorticoids induce many of the enzymes of gluconeogenesis. Androgens and oestrogens cause profound changes in the enzyme profile of the gonads and other tissues at puberty, and progesterone has similar massive effects during pregnancy. There was a long gap between the identification of this basic effect of steroid hormones and the development of any detailed understanding of the mechanism of action. First, a good description of the processes of transcription and translation in protein synthesis was necessary. It is difficult to characterize the regulation of a process without first understanding the basic mechanisms involved. A second problem was the difficulty of determining the location of the hormone inside the cell. This was overcome by the introduction of tritium-labelled hormones in the mid-1960s.

11B Structure of steroid hormones

An example of a steroid hormone, corticosterone, is shown in Fig. 11.1. All the hormones have the same basic four-ring structure. Hormones require a highly specific three-dimensional structure so that the recognition site on the receptor can distinguish between one hormone and another. In the case of polypeptide hormones the specificity lies in the amino acid sequence, and in the case of larger polypeptides in the three-dimensional structure as well. It is less immediately obvious how steroid receptors discriminate between one steroid and another, the molecules being superficially very similar. However, the ring system is approxi-

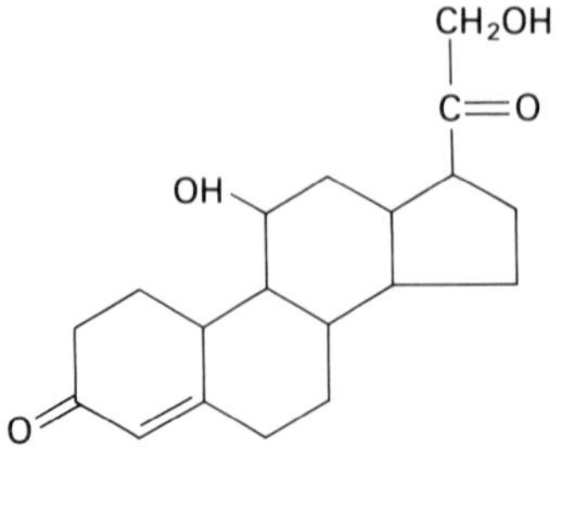

Fig. 11.1. Structure of corticosterone.

mately planar and there are numerous possibilities of substitution either above or below the ring. This leads to a very complex stereochemistry which provides the necessary discrimination.

11C Mechanism of action

A hypothesis for the mechanism of action of steroid hormones was developed in the late 1960s. It followed from experiments in which the time course of accumulation of ^{3}H-labelled hormone into tissues was followed. Steroid hormones were found to enter all cells freely but to accumulate to a concentration higher than the medium only in target cells. The accumulation was found first in the cytoplasm and then in the nucleus. This type of study led to the model summerized in Fig. 11.2. The sequence of events suggested was as follows.

1 The hormone enters the cell.
2 The hormone binds to a cytoplasmic receptor protein. The receptors

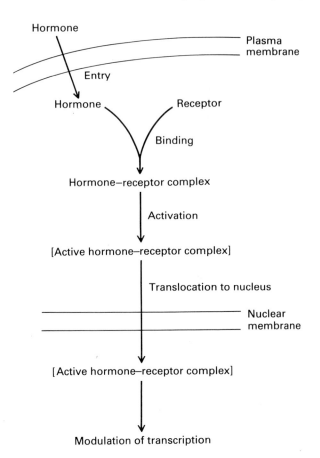

Fig. 11.2. Model for the mechanism of action of steroid hormones based on the time of accumulation of ^{3}H-labelled hormone in cytoplasm and nucleus.

are present only in target cells and it is this binding which is responsible for the accumulation of the hormone in the target tissue.

3   There is an activation process dependent upon the binding of the hormone which leads to a change in the cytoplasmic receptor.

4   As a consequence of activation, the receptor acquires an affinity for, and accumulates in, the nucleus.

5   In the nucleus, the activated hormone–receptor complex interacts with DNA in some way to activate the transcription of specific genes.

6   Finally, there must be a mechanism for switching the effect off.

Subsequent work has led to a much more detailed understanding of the mechanism of action of steroid hormones, but in general the model has held up well. In the following sections each step will be considered in detail.

11D Entry into the cell

Steroid hormones are small, highly lipophilic molecules which might be expected to pass freely across biological membranes. Early studies with ^{3}H-labelled oestradiol found that the hormone became rapidly associated with all cell types and that there appeared to be no discrimination between target cells and non-target cells. This led to the suggestion that steroid hormones pass freely across the plasma membrane.

More recently the possibility that some specific transport mechanism exists has been re-examined. Steroids do not enter the cell at 0°C and this was taken to suggest that there is a specific transport protein which, like most membrane proteins, has a very high Q_{10} and therefore little activity at low temperature. However, the loss of entry could result from the reduction in mobility of the membrane lipid, caused by the reduction in temperature, resulting in a general change in permeability to steroids. It has also been suggested that the uptake of steroids into cells follows saturation kinetics. This is a critical criterion for establishing the existence of a carrier protein, since a simple diffusion process has a linear relationship of rate of uptake with concentration over a wide concentration range (Fig. 11.3). With steroids, such studies are always very difficult since the level of non-specific binding to cells is invariably very high due to the lipophilic nature of the molecule. As a result it is difficult to distinguish between binding and entry except when there is substantial accumulation, as is observed in target tissues.

The possibility that there is a carrier-mediated mechanism for the entry of steroid hormones into the cell cannot be excluded, but the evidence in its favour is not very convincing. There is no evidence that such a mechanism has any significant role to play in the recognition of

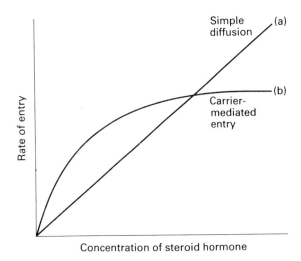

Fig. 11.3. Relationship of rate of entry to concentration. Entry by simple diffusion across the membrane shows a linear relationship with hormone concentration (a). For a carrier-mediated process saturation kinetics would be expected (b).

the hormone by target cells, or in providing specificity in the target-cell response.

11E Acute effects of steroid hormones

Steroid hormones are known to cause rapid responses in target cells which are far too fast to be explained by alterations in rates of enzyme synthesis. Such effects are poorly characterized. In most cases the cellular response is not well defined and there is little information about possible mechanisms. Some of the best characterized effects have been described in neural tissue where steroid hormones have been shown to affect neuronal discharge rates in some areas of the brain both directly and by modulating the response to neurotransmitters. Steroids have also been shown to affect the number of neurotransmitter receptors on the cell surface very rapidly. Acute effects in non-neural tissue almost certainly occur but are less well described perhaps because they tend to be obscured by much larger responses to the classical acute hormones. It seems likely that such acute effects require a cell-surface receptor mechanism.

As described in the previous section, there is some evidence for saturable binding sites on the cell surface. It is possible that these are the receptors which mediate acute effects rather than steroid transport proteins. There is no information at all about the course of events following the binding of a steroid hormone to cell-surface receptors, assuming that they do exist. The whole problem of the nature of acute affects of steroids requires clarification.

11F Modification

After entry into the cell the hormone may be modified. A well characterized example is the case of testosterone which is reduced by testosterone 5α-reductase to give dihydrotestosterone. (Fig. 11.4). In the mature male, certain tissues of the genitals respond to testosterone while others respond to dihydrotestosterone. The 5α-reductase is only present in cells which respond to dihydrotestosterone, so that in effect in these tissues a different hormone is generated from the original hormone within the cell. The presence of the 5α-reductase enzyme is itself a developmental change which occurs at puberty.

11G The steroid hormone receptor

11G1 Identification of receptors: binding studies

As with measurement of cell-surface binding, the measurement of binding of steroids to specific intracellular sites is complicated by high levels of non-specific binding. However, it is possible to identify unequivocally sites which have the properties expected of a hormone

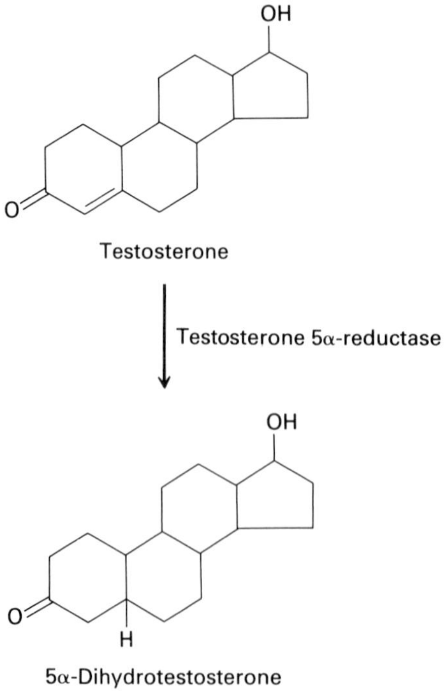

Fig. 11.4. Intracellular reduction of testosterone.

receptor, that is, high affinity and a limited total number of sites. Typically the receptors have a dissociation constant for the hormone in the range $10^{-9} - 10^{-10}$ mol/l and a total concentration of binding sites in the cytoplasm of 10^{-8} mol/l. Steroid hormones often affect a wide variety of different tissues producing different responses in each tissue. This raises the possibility that different classes of receptor are involved in the different responses. However, this does not appear to be the case. A single genetic lesion leading to a non-functional testosterone receptor leads to the loss of all the effects of testosterone.

The use of analogues of steroid hormones has been an important method for the identification of specific receptor sites. Drugs which will substitute for the hormone should also compete with the hormone for binding to the receptor. There should also be agreement between the relative effectiveness of different drugs in causing a physiological response and their relative binding affinities for the receptor. Less effective drugs should show weaker binding. Steroid analogues which block the action of the hormone should also interfere with binding. The use of such analogues is an important way of establishing that proteins which have high-affinity binding sites for steroid hormones are also functionally significant.

Two major technical problems have made it difficult to characterize the steroid receptor. One is the high level of non-specific binding. A second more serious problem is the difficulty of maintaining the integrity of the receptor during extraction of the cells and purification. The state of aggregation of the receptor is affected by the ionic composition of the extraction media and several different high-affinity binding species can be detected. A number of proteolytic fragments are also observed. There are two important questions.

1 Which of the many species which can be detected represent the normally functioning state *in vivo*?
2 Where in the cell is the receptor located?

The first method used to identify receptors was to fractionate cytoplasmic extracts by sucrose density centrifugation and to determine the position on the gradient of the steroid-hormone binding capacity. Depending upon the protein concentration of the extract and upon the salt content of the medium, binding proteins could be found with sedimentation coefficients of $4s$, $6s$ or $8s$. The larger sizes were favoured by high protein concentration and low salt.

A smaller species of binding protein sedimenting at $2-3s$ was also observed. The proportion of the binding in this state could be increased by mild proteolytic treatment of cytosol extracts and it appears that these small proteins known as meroreceptors are generated as a result of endogenous protease activity. The picture was further complicated by the observation that steroid-binding fractions isolated from the nucleus had a sedimentation coefficient of 5s.

11G2 Functional role of the different binding proteins: receptor activation

All the different fractions represent a relatively small number of high-affinity binding sites for the steroid hormone and are accordingly possible candidates for the functional receptor. The ideal approach to establish further the role of a binding protein is to purify it and examine the effects of the pure protein on the physiological response. However, successful purification of steroid-hormone receptors has only been achieved quite recently and earlier studies depended upon more indirect approaches.

Three types of observation suggested that the 4s, 6s and 8s cytoplasmic binding fractions represent functional receptors.

1 For a particular hormone, they were found only in target-tissue cells.

2 Agents which antagonize the action of steroid hormones also inhibited binding to the 4s, 6s and 8s fractions. For example, the drug nafoxidine, which blocks oestradiol action, also blocked binding to the putative receptors.

3 The number of binding sites found in the cytoplasm decreased as protein-bound hormone appeared in the nucleus. This suggests that the cytoplasmic receptors act as precursors for the nuclear 5s hormone–receptor complex. A number of observations supported this idea.

Nuclei isolated from cells which had not been exposed to steroid hormones contained very little hormone-binding activity. In intact cells, analogues which block binding to the cytoplasmic 4s, 6s and 8s components also prevented the appearance of 5s-bound hormone in the nucleus. The cytoplasmic 4s, 6s and 8s binding proteins will bind hormone at 0°C. The appearance of the 5s component in the nucleus, however, required a temperature in excess of 25°C. Thus the minimum requirements for the accumulation of the 5s hormone–receptor complex in the nucleus are the presence of hormone, cytoplasm, nuclei, and a temperature in excess of 25°C. It follows that one of the cytoplasmic components, 4s, 6s or 8s, acts as a precursor for the nuclear 5s hormone–receptor complex and that there is an activation process converting the one to the other. The rate of appearance of the 5s component depended upon the concentration of the cytoplasmic extract, and had higher than first-order kinetics suggesting the involvement of more than one component. Taken together with the marked temperature dependence, this suggests that the activation involves some sort of enzymic conversion. The activation process leads to a marked increase in affinity for the nucleus and to the acquisition of the ability to affect messenger RNA synthesis.

The early, relatively crude cell-fractionation experiments led to a fairly well-defined model (Fig. 11.5). However, many questions remained unresolved. It was not clear which of the several cytoplasmic receptor species represent the normal functional state. The need for low ionic strength to detect the larger species suggests that the 4s species is the normal state. On the other hand, the larger species are favoured by high protein concentrations as would occur in the cytoplasm of the intact cell. The nature of the transition of the cytoplasmic receptor to the activated 5s receptor was also not at all clear. If the 4s receptor is the usual functional state, a change to 5s implies either a major conformational

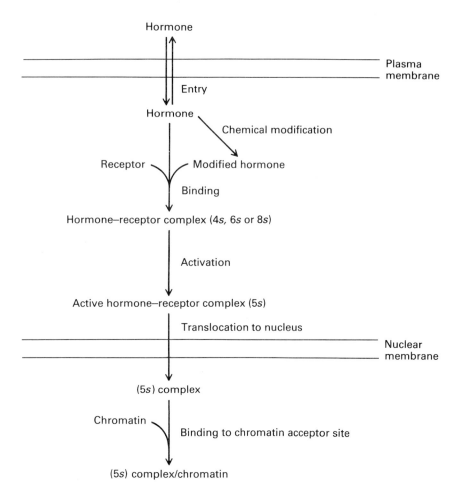

Fig. 11.5. Model of steroid hormone action based on density-gradient determination of receptor size and location.

change or the association of another component. If the larger components are the usual functional state then the dissociation of subunit or removal of a component by proteolytic cleavage are possibilities. Another question to consider is the transfer of receptor from the cytoplasm to the nucleus. If the initial binding capacity is found only in the cytoplasm then it follows that the receptors are excluded from the nucleus; on activation they acquire the ability to enter the nucleus. This suggests that there is some sort of controlled transfer process. Some of these questions have been resolved by the purification of receptors.

11H Pure receptor studies

11H1 Properties and structure

The progesterone receptor of chick oviduct has been extensively studied. The dissociation constant for progesterone is between 4×10^{-10} and 5×10^{-9} mol/l. Different figures have been obtained depending upon whether the constant is determined by equilibrium binding or from the rate constants of association and dissociation. This indicates a complex binding mechanism. There are between 10 000 and 40 000 copies of the receptor per cell. The binding specificity of the receptor for different analogues of progesterone correlates well with the potency of the same analogues in producing the physiological response. The retention of this order of binding affinity was a useful test of the various stages of purification.

As in other systems, sucrose density-gradient fractionation yielded binding components of $4s$, $6s$ and $8s$, high ionic strength favouring the $4s$ component. A more careful analysis showed that the $4s$ component was not homogeneous. It could be separated by ion-exchange chromatography into two fractions: A with a sedimentation coefficient of $3.6s$ and B with a sedimentation coefficient of $4.5s$. Both bound progesterone with high affinity. A has a molecular weight of 75 000 and B of 117 000, and both are highly asymmetrical in shape. High salt treatment of the $6s$ fraction yields equal amounts of A and B and it is possible to reconstitute the $6s$ component from purified A and B. The possibility that A was derived from B by proteolysis was considered. The peptide maps of tryptic digests of the two proteins show many differences indicating that they are distinct proteins rather than one being derived from the other. These studies suggest that the $6s$ component is the normal state of the receptor and contains one $4sA$ subunit and one $4sB$ subunit. The $8s$ form appears to consist of non-specific aggregates of larger size.

11H2 Nature of the activation and accumulation in the nucleus of steroid receptors

Earlier studies on steroid hormone action suggested that the binding of hormone and activation of the receptor is an exclusively cytoplasmic process. The activation also appeared to involve a substantial change in the physical nature of the receptor reflected in an increase in sedimentation coefficient from $4s$ to $5s$.

More recent experiments to determine the localization of progesterone binding within target cells suggest that the binding capacity is evenly distributed throughout the cell and occurs in both the cytoplasm and the nucleus. There are far more receptors in the cytoplasm as a consequence of its greater volume but no difference in concentration. The accumulation of receptor in the nucleus after binding is simply a consequence of increased binding affinity and there is no need to postulate either a translocation process or that the nucleus has the ability to exclude free receptor but not activated hormone−receptor complex.

When activated progesterone−receptor complex was extracted from the nucleus it was found to have a sedimentation coefficient of $4s$ and to consist of equal amounts of the A and B forms, again suggesting that the $6s$ form is the initial binding state. There was no evidence of any major physical change relative to the hormone-free receptor but minor physical changes such as protein phosphorylation would not have been detected.

These observations are consistent with a much simpler model. The hormone binds to the $6s$ form of the receptor which consists of one A subunit and one B subunit. As a consequence of binding, the affinity of the receptor for binding sites in the nucleus increases, leading to accumulation in the nucleus. The model fails to explain the temperature-dependent cytoplasmic activation process which has complex kinetics. It has been suggested that this represents the reversal of a non-physiological state resulting from the extraction procedure.

Recent studies on the oestrogen receptor suggest that it may be located in the nucleus. Two approaches have been used. In one series of experiments, monoclonal antibodies to the oestrogen receptor were used to detect its presence in intact permeabilized cells. The other group made use of a very gentle procedure to separate the cytoplasm from the nucleus. In both cases the receptor appeared to be located in the nucleus, even in the absence of the hormone. It is likely that, in the absence of hormone, the receptors are either evenly distributed throughout the cell or located in the nucleus. Experiments have employed different hormones and different cell types so it is not as yet possible to come to a firm conclusion about which is the correct location. The possibility that different steroid receptors are located differently in the cell cannot be excluded.

11H3 Nuclear binding

The binding of the hormone−receptor complex to the nucleus shows tissue specificity, in that there is more binding to target-cell nuclei than to non-target cell nuclei. This observation led to the suggestion that the nucleus contains specific acceptor sites to bind the hormone−receptor complex. There is no obvious requirement for specificity at this level, the need for specificity having been satisfied by the presence or absence of the receptor. However, such specificity does occur.

Three classes of binding site can be identified. In an individual nucleus there are about 1000 sites of high affinity, 10 000 of intermediate affinity and 100 000 of low affinity. Only the high-affinity sites approach saturation at physiological hormone concentrations and it seems likely that the other sites are of no physiological significance. In fact the lower-affinity sites cannot be detected at physiological concentrations of KCl.

11H4 Binding to chromatin

The binding of progesterone receptors to chromatin has been examined using both crude cytosol extracts and purified receptor. There were no obvious changes in the binding properties during the process of purification. Binding is tissue specific. As might be expected, non-target cell cytosol does not give rise to any binding of labelled progesterone to target-cell chromatin due to the lack of receptors. Target-cell cytosol in the presence of hormone does give rise to some binding to non-target cell chromatin, but binding to chromatin from a target cell is much more extensive.

In experiments with purified receptor it was found that only the $4s$B receptor or the $6s$ dimer bound to chromatin. The $4s$A subunit had no affinity for chromatin at all, but did bind to DNA. Binding to DNA was not tissue specific. Since all the cells of an organism contain the same DNA this is not surprising. Association of the $4s$B receptor with chromatin led to an increase in the already tight binding affinity for the hormone, resulting in an increase in the half-time for dissociation from 25 to 100 minutes.

11H5 The nuclear acceptor site

The next question to consider is the nature of the nuclear acceptor site in chromatin. Chromatin consists of DNA, histones and non-histone proteins (sometimes called acid proteins to distinguish them from the highly basic histones). As mentioned above, the 4sA receptor binds to exposed DNA but not to intact chromatin and not to any of the proteins in chromatin. The binding capacity for the B receptor was found to be in the non-histone proteins. Histones have no binding capacity for any form of the receptor, which is not surprising since there are only five different histones of which four show no tissue specificity. A cell may respond to many more than five different steroid hormones. Non-histone proteins, on the other hand, are numerous; there may be as many as 500 different proteins in a single tissue. They are also thought to contain the factors which are responsible for regulating DNA transcription.

Non-histone proteins can be subdivided into three fractions, AP1, AP2 and AP3, on the basis of the physical treatments needed to extract them from the chromatin. The acceptor sites appear to reside in the AP3 fraction which constitutes about fifty per cent of the total non-histone protein. If AP3 proteins from a target cell are recombined to form chromatin with histones, DNA, and AP1 and AP2 proteins from non-target cells, the chromatin has a high binding capacity for the B receptor. If the reverse is done, and AP3 proteins from non-target cells are reconstituted to form chromatin with the other components derived from target cells, then the binding capacity is reduced to about twenty five per cent. This is about the same as the amount of binding seen in chromatin from non-target cells.

It might seem simpler to measure the binding of the AP3 fraction directly, but isolated AP3 proteins do not bind 4sB receptors and need to be combined with DNA and histones for the binding activity to be expressed. Unfortunately this means that it will be very difficult to purify the individual protein, or proteins, which contain the acceptor site since during isolation it loses its binding activity and hence the means of detecting its presence. It is not known if the receptor for a particular steroid hormone has a single type of acceptor protein for all the genes it activates, or if there are many different acceptor proteins, the most extreme case being a different acceptor for each different gene.

In summary, 4sB receptors and 6sAB dimers bind to chromatin but not to DNA; 4sA receptors bind to DNA but not to chromatin. If the 4sA hormone–receptor complex has a physiological role, this gives further support to the view that the 6s dimer is the normal functional state. This must be the case since the A subunit has affinity for only the nuclear acceptor site when it is part of the 6s complex.

11J The nature of the nuclear response

One of the earliest observations relating to steroid hormones was that they increase the rate of synthesis of specific proteins. Later it was shown that the rate of messenger RNA synthesis was increased, suggesting that the control is at the level of transcription. More recently, purified receptors have been shown to increase the rate of synthesis of messenger RNA from chromatin in a cell-free system. This effect was confined to target-tissue chromatin, and showed saturation kinetics with respect to the concentration of the hormone−receptor complex. The time course of onset of mRNA synthesis was consistent with the time course of hormone−receptor binding to the chromatin. Different components of the hormone receptor were tested for their ability to promote messenger RNA synthesis from target-cell chromatin. The 4sB receptor was ineffective, the 6s dimer was highly effective, and the 4sA receptor was effective but required much higher concentrations than the 6s dimer. Taking account of the known binding properties of the different components, this suggests that the intact 6s receptor is needed for binding to chromatin using the affinity of the B subunit for the acceptor site in the AP3 protein fraction. The A subunit then promotes the response, interacting with DNA to affect the rate of transcription.

11J1 Binding of the A subunit to DNA

The 4sA subunit has the properties of a DNA helix-destabilizing protein in that it binds to both single- and double-stranded DNA but with a marked preference for single-stranded DNA. Until very recently there was no evidence that the 4sA subunit could recognize specific sequences of DNA, implying that specificity of activation of gene transcription resulted from the interaction of the 4sB subunit with the protein acceptor site in chromatin. The binding of the 4sB subunit to the acceptor protein would lead to the exposure of the DNA to the 4sA subunit which would then activate the transcription of the gene. This is an attractively simple model but it leaves many questions unanswered.

The interaction of steroid hormones with a target cell usually leads to many genes being expressed, and a figure of 1000 for the number of acceptor sites is quite reasonable bearing in mind that this is equivalent to 500 different genes. Also, the figure is probably an overestimate of the number of functional acceptor sites since substantial binding occurs

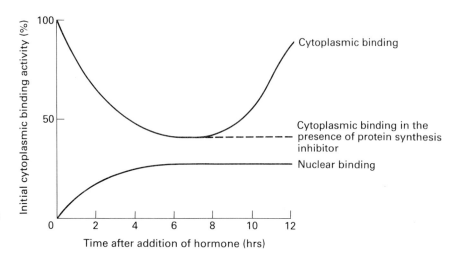

Fig. 11.6. Changes in cytoplasmic and nuclear steroid hormone-binding activity after a large dose of hormone.

in non-target cell chromatin. However, the extent to which different genes are expressed varies greatly and an explanation is needed for this. The model also implies that the acceptor protein or proteins are able to recognize specific sequences of DNA in order to locate themselves near to the appropriate gene.

More recent studies suggest that the receptor can recognize specific sequences of DNA. Glucocorticoid receptors have been shown to bind to a DNA sequence close to the promoter region of responsive genes, so there appears to be specificity of binding both of the B subunit to the protein-acceptor site and of the A subunit to DNA.

Steroid hormone receptors also appear to increase the association of target genes with the nuclear matrix. The nuclear matrix is a protein fibrillar structure which is thought to be involved in transcription and to be associated with genes which are being actively transcribed. Part of the chromatin in the cell is associated with the nuclear matrix as is heterogeneous mRNA, the initial product of transcription.

11K Reversal of the action of steroid hormones

The mechanism by which the activation of protein synthesis by steroid hormones is switched off is poorly characterized. It appears to involve the proteolytic cleavage of the hormone receptor rather than the release or degradation of the steroid hormone itself. This idea is supported by a number of observations. First, there is no evidence for the breakdown of steroid hormones in their target cells. Second, if a target cell is exposed

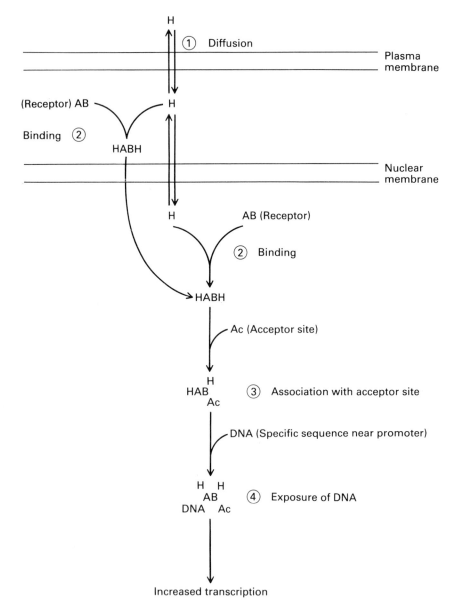

Fig. 11.7. Model of steroid hormone action. H=hormone, AB=receptor, Ac=acceptor site in chromatin. The model assumes a random distribution of receptors between the cytoplasm and the nucleus so that binding of the hormone may occur in either location.

to a high concentration of a steroid hormone, the receptor activity in the cytoplasm can become depleted by as much as sixty per cent. Substantially less receptor appears in the nucleus than is lost from the cytoplasm. The reappearance of receptor in the cytoplasm takes about twelve hours and is blocked by the addition of inhibitors of protein synthesis (Fig. 11.6). This suggests that after binding of the hormone the receptor becomes susceptible to proteolysis, and that this is the means of reversing the response. The small meroreceptors have been suggested to represent an intermediate in this proteolytic breakdown.

11L Conclusions

The processes regulated by steroid hormones are very complex so it is not surprising that we do not have a complete picture of the mechanism of action. The current model is summarized in Fig. 11.7. As progress is made towards a detailed understanding of gene expresion, the mechanisms by which steroid hormones modulate the process should become clearer.

Suggestions for further reading

Milgrom E. (1981) Activation of steroid hormone receptor complexes. In *Biochemical Actions of Hormones* (Ed. by G. Litwack), vol. 8, pp. 466–493. Academic Press.

Schrader W.T., Birnbaumer M.E., Hughs M.M., Weigel N.L., Grady W.W. & O'Malley B.W. (1981) Studies on the structure and function of the chicken progesterone receptor. *Recent Progress In Hormone Research*, **37**, 583–635.

International Review of Cytology. Supplement 15, p.1.

Chapter 12 Growth factors

12A Introduction

The existence of polypeptide growth factors was first recognized in the 1950s and widespread interest in the subject dates from about 1970. The first to be discovered was nerve growth factor. It was found that transplanting a mouse sarcoma into chicken embryos led to an increase in the outgrowth of nerve fibres from adjacent spinal ganglia, and also to an increase in the size of the ganglia. The effect was found to be caused by a protein released from the tumour which was given the name nerve growth factor (NGF). The amounts of NGF produced were small so a large number of mouse tissues were examined to find a better source. The submaxillary gland of the male mouse turned out to contain far more of the factor than any other tissue and this led to the accidental discovery of a second growth factor, epidermal growth factor (EGF). Crude extracts of mouse submaxillary gland stimulated the growth of epidermal and endothelial tissues in new-born mice. A second peptide EGF, distinct from NGF, was found to be responsible.

Over the last ten years or so, many other growth factors have been identified (Table 12.1). It had been recognized for a long time that normal mammalian cells would not survive in culture unless whole serum was present. Attempts to characterize the essential components

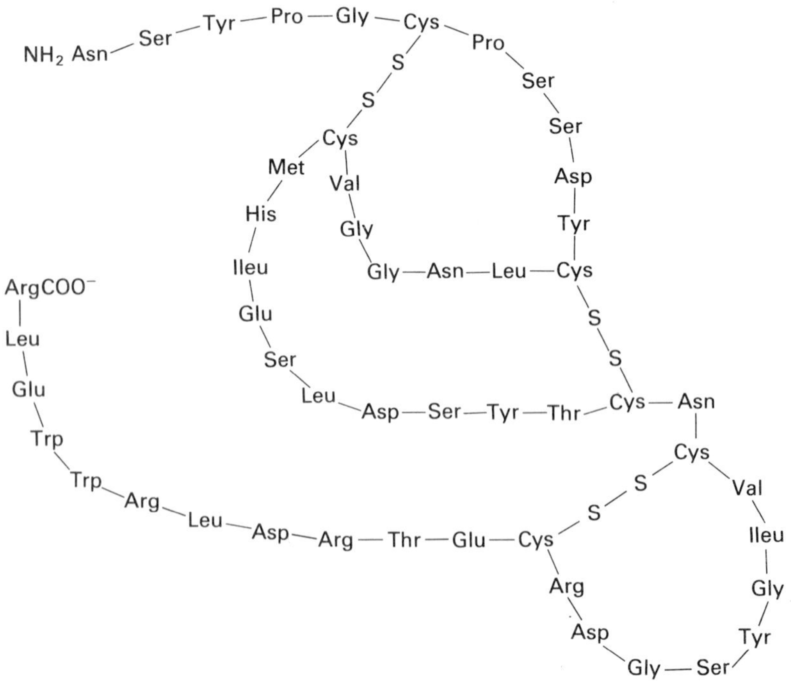

Fig. 12.1. Amino acid sequence of mouse EGF.

Table 12.1. Peptide growth factors

Factor	Source
Epidermal growth factor (EGF)	Mouse submaxillary gland
Nerve growth factor (NGF)	Mouse submaxilliary gland
Fibroblast growth factor (FGF)	Pituitary
Insulin-like growth factors (IGF 1) (IGF 2) Somatomedin A (SMA) Somatomedin C (SMC)	Serum
Platelet-derived growth factor (PDGF)	Platelets
Fibroblast-derived growth factor (FDGF)	Fibroblasts
T-cell growth factor IL-2	Lymphocytes

led to the identification of a group of polypeptide growth factors in serum, the insulin-like growth factors (IGF). Other growth factors have been found to be released by particular cell types such as platelet-derived growth factor (PDGF) and fibroblast-derived growth factor (FDGF). A factor which stimulates fibroblast growth (FGF) was found in the pituitary. NGF and EGF have been studied for the longest period of time and are available in relatively large amounts. As a result they are the best characterized and the rest of this chapter will concentrate on the action of NGF and EGF.

12B Source and structure of EGF and NGF

Both EGF and NGF are usually isolated from the submaxillary gland of the adult male mouse which is by far the richest source. The physiological reasons for this are obscure since neither immature male mice nor female mice at any stage of development have such high levels in the submaxillary gland. Other rodents also lack high levels of the growth factors in the gland.

Removal of the gland has no effect on nerve function or growth and NGF appears to be secreted into the saliva rather than into the circulation. Many cell types have been shown to produce NGF in culture including glioma cells, neuroblastoma cells, skeletal muscle, fibroblasts and the non-neural cells of dorsal root ganglia. This might be taken to suggest that NGF is either produced by tissues which are innervated or by satellite cells in the ganglia. However, there is no evidence for either proposal *in vivo* and it is not clear what the physiological source of NGF is.

The concentration of EGF in serum is about 10^{-10} mol/l, similar to most hormones. The mouse submaxillary gland does secrete EGF into serum and this is stimulated by α-adrenergic agonists. However, it seems likely that other sources are physiologically significant but these are not known. A similar factor is found in human urine, which inhibits gastric acid secretion in the stomach. This peptide, known as urogastrone, has 37 out of 53 amino acid residues identical with mouse EGF and the disulphide bridges are all in the same positions. It may well be identical to human EGF, but again the tissue source is not known.

12B1 Structure of EGF

The active form of mouse EGF is a single polypeptide chain of 53 residues with a molecular weight of about 6000. It is quite acidic with an isoelectric point of 4.6. There are three disulphide bridges which must be intact for biological activity. The sequence of mouse EGF is known (Fig. 12.1). A high molecular weight form of EGF can be isolated from mouse submaxillary gland, which consists of two molecules of EGF, two molecules of an EGF-binding protein and two Zn^{2+} ions (Fig. 12.2). The binding protein is an arginine-specific protease, and since the C-terminal amino acid of EGF is arginine it is likely that the binding protein processes a precursor of EGF by removing a C-terminal peptide, and that a small proportion of the EGF remains bound to the protease.

12B2 Structure of NGF

NGF is isolated from the mouse submaxillary gland as a large complex (7s), with a molecular weight of 130 000. It consists of three types of subunit, two α, one β and two γ, and also contains two Zn^{2+} ions. The β subunit is responsible for the biological effects of NGF. It is a dimer, with a molecular weight of 26 000, made up of two identical subunits each with 118 amino acid residues. The isoelectric point is 9.3. The association between the subunits is very tight with a dissociation constant of 10^{-13} mol/l suggesting that it probably functions **as a dimer** *in vivo*. Like EGF, the C-terminal residue is arginine.

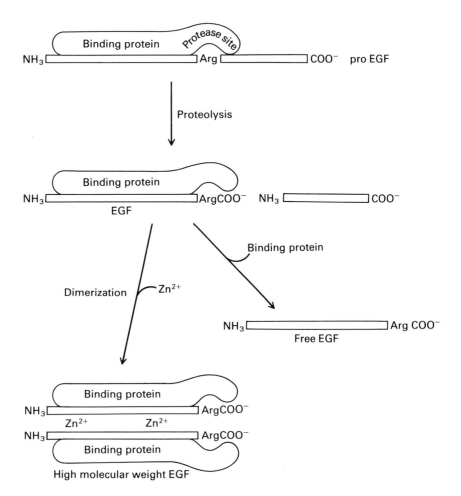

Fig. 12.2. Processing of EGF. After proteolysis, EGF may either form a dimer together with the binding protein and Zn^{2+}, or release the binding protein to form free EGF. Free EGF is the form found in the circulation.

The α subunit has a molecular weight of 26 000 and is acidic with an isoelectric point of 4.3. No biological activity has been found other than its ability to form part of the complex. The γ subunit also has a molecular weight of 26 000 and has been found to be an arginine-specific protease. There are obvious similarities to the situation with EGF. It has been suggested that the γ subunit binds to an NGF precursor and generates the β subunit by removal of a C-terminal peptide. Unlike EGF, a third protein, the α subunit, then binds and stabilizes the complex as a storage form (Fig. 12.3). NGF is secreted into saliva as the high molecular weight complex, but it is not known which form is released into the circulation. The high molecular weight complex has no biological activity.

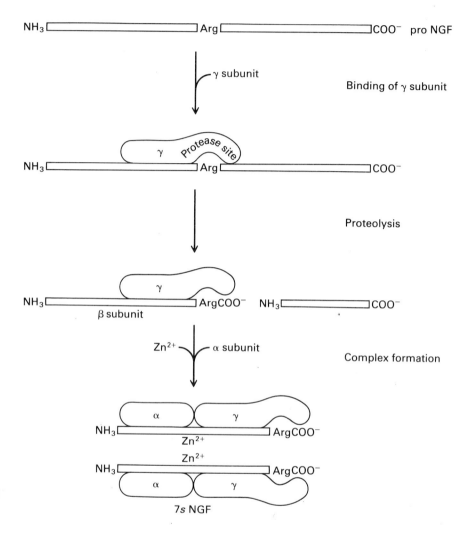

Fig. 12.3. Formation of 7s NGF from pro NGF.

12C Physiological effects of growth factors

The characteristic effect of growth factors is the stimulation of cell growth and proliferation. Different growth factors have different target cells but it seems likely that all cell types respond to one or more different growth factors.

12C1 Physiological effects of EGF

In vivo injection of EGF into neonatal animals causes accelerated differentiation and proliferation of epidermal and epithelial tissue and also of liver cells. Changes in the tissues include the induction of the enzyme ornithine decarboxylase and an increase in polyamine concentration, which are both changes associated with increased rates of cell division.

In cultures of cells derived from target tissues, EGF increases the rate of cell growth and division. As might be expected there is a general increase in the biosynthesis of DNA, RNA and proteins. There are changes in the plasma membrane, and the synthesis and secretion of prostaglandins increase.

There are also rapid general anabolic responses similar to the effects of insulin. Transport of glucose, amino acids, and the purine and pyrimidine precursors of nucleic acids is increased. The activity of the Na^+/K^+-ATPase is also increased.

12C2 Physiological effects of NGF

NGF controls the growth of sensory and sympathetic ganglia. If antibodies to NGF are injected into young rats or mice (a few days old) there is almost complete destruction of many of the sympathetic ganglia and of some sensory ganglia. Injection of anti-NGF into embryos leads to destruction of virtually all of the sympathetic ganglia. In adult animals, antibodies to NGF do not cause irreversible damage to the sympathetic ganglia but do impair function. It seems that NGF is required for sympathetic nerve growth in neonatal animals and for the maintenance of nerve function in adults.

NGF enhances cell division in the non-neural cells of dorsal root ganglia. This can be demonstrated both *in vivo* and with ganglia in culture. There is no increase in the number of neural cells but their volume is increased. It seems that the effect on the non-neural cells is mediated by the nerve cells since NGF has no effect unless intact nerve cells are present.

The acute effects of NGF at the cellular level are similar to those of EGF. Glucose transport, amino acid transport and nucleoside transport are all increased and the Na^+/K^+-ATPase is activated. The turnover of phosphatidylinositol is increased. Longer-term effects are more difficult to study in that nerve cells will not survive in culture for prolonged periods in the absence of NGF. The control cultures enter a phase of irreversible degeneration and become rather meaningless as controls. However, there is a general increase in protein synthesis, RNA synthesis and, in non-neural cells, DNA synthesis. More specific effects include the induction of certain enzymes, in particular, tyrosine hydroxylase and dopamine β-hydroxylase which are key enzymes in the pathway of catecholamine biosynthesis.

12D Mechanism of action of EGF

12D1 The mitogenic response

The usual way to study the effect of EGF and other growth factors is to measure the incorporation of [^{3}H]-thymidine into DNA in cultured cells. A variety of cell types have been used but the most common are cultures of fibroblasts or of cell lines derived from fibroblasts. The 3T3 cell line has been particularly widely used. The usual approach is to employ cultures of quiescent cells, that is cells which are arrested in the G_0/G_1 phase of the cell cycle (Fig. 12.4). In other words, the cells are not synthesizing DNA (S phase), or progressing towards mitosis and cell division (M phase) (Fig. 12.4). This tends to happen naturally when the cell culture grows to confluence as a result of contact inhibition. The ability of exogenous factors to stimulate such cells to move into S phase and initiate DNA synthesis can then be measured.

12D2 Synergistic effects

Early work with EGF and with other growth factors employed culture media which contained serum to maintain the culture. More recently it has been possible to culture many cell lines in serum-free media, which makes it possible to define all the growth effectors added. In most cases the addition of two different growth-promoting effectors has a highly synergistic effect on DNA synthesis. As an example, the effects of EGF and insulin, alone and together, are shown in Table 12.2. In many cases a single factor gives virtually no response on its own. A number of effectors have been found to be synergistic with EGF including insulin,

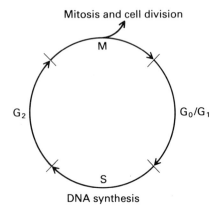

Fig. 12.4. The cell cycle. For the determination of growth-promoting effects, cells in the quiescent G_0/G_1 phase are used. DNA synthesis is determined by the incorporation of [^{3}H]-thymidine and this gives a measure of the ability of a growth factor to promote the transition to S phase.

Growth factors

Table 12.2. Relative rates of DNA synthesis in 3T3 cells

No additions	1
EGF	2
Insulin	8
Insulin + EGF	110

fibroblast-derived growth factor (FDGF), the pituitary hormone vasopressin, and phorbol esters which promote malignant transformation. The most powerful phorbol ester is TPA (12-O-tetradecanoylphorbol-13-acetate) (Fig. 12.5). The need for two independent effectors to promote the mitogenic response complicates studies on the mechanism of action. In particular, there is the problem of deciding which effector is responsible for generating a particular intracellular signal.

12D3 The EGF receptor

Saturable binding of EGF is found in many types of cell, which are not all of epidermal origin (Table 12.3). The dissociation constant is between 10^{-10} and 10^{-9} mol/l. The receptor density varies markedly between cell types: typically there are around 10^4 receptors per cell but in epidermal carcinomas there may be over 10^6 per cell. The receptor was first identified in 3T3 cells with a photoreactive analogue of EGF labelled with [^{125}I]-iodine. A single protein was labelled, which had a molecular weight of 184 000, by SDS gel electrophoresis. The receptor has now been identified in a number of different tissues and there is broad agreement about its molecular weight. It has also been purified from a human carcinoma cell line which has an abnormally high receptor density. The pure receptor has a molecular weight of 170 000 but is easily proteolytically cleaved to give a 150 000 form which still binds EGF. The receptor is a glycoprotein. EGF binding is inhibited by lectins, and incubation of cells with tunicamycin, an inhibitor of protein glycosylation, leads to a gradual loss of EGF binding activity.

12D4 Receptor internalization

At 37°C, [^{125}I]-labelled EGF is rapidly broken down by cells yielding mono-iodotyrosine, while at the same time the number of receptors on the cell surface is reduced. This effect is blocked by inhibitors of lysosomal function. The breakdown process has been followed using EGF labelled in a number of different ways. It appears that the hormone

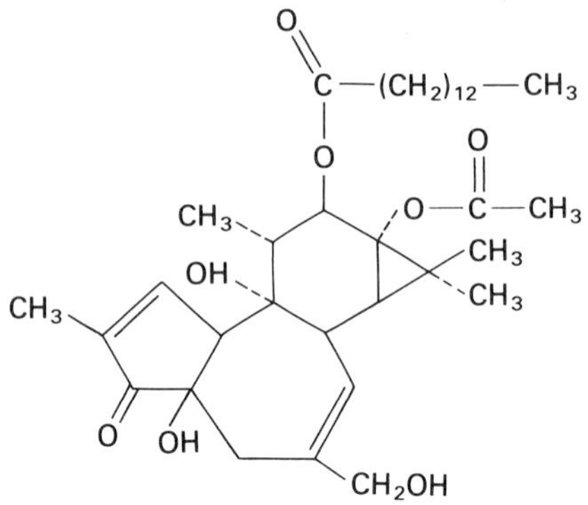

Fig. 12.5. Structure of TPA (12-0-tetradecanoylphorbol-13-acetate).

binds to receptor sites which are randomly dispersed on the cell surface. This is followed by aggregation, and internalization of the receptors, together with the bound EGF. Aggregation and internalization can also be promoted by antibodies to the receptor. Both the receptor and EGF are broken down inside the cell. If the receptor is labelled, as well as EGF, a number of proteolytic products can be identified, associated with the lysosomal fraction of the cell. EGF itself is broken down about five times as fast as the receptor suggesting that most of the internalized receptors are recycled back to the membrane. Most hormones show a reduction in receptor number on the cell surface as a result of prolonged exposure to the hormone. In most cases the function seems to be desensitization or down regulation, but in the case of growth factors the possibility of a role in the cellular response must be considered (see Section 12E1).

Table 12.3. Cell types with EGF receptors

Corneal cells

Fibroblasts

Lens cells

Glial cells

Epidermal carcinoma cells

3T3 cells

Granulosa cells

Vascular endothelial cells

Placental membrane

12D5 Tyrosine kinase activity of the receptor

Purification of the EGF receptor by affinity chromatography has revealed that the receptor has a tyrosine-specific protein kinase activity. The activity co-purifies with the receptor and is a function of the same protein. The receptor phosphorylates itself and also other proteins. Mild proteolysis of the receptor, reducing its molecular weight from 170 000 to 150 000, leads to an increase in its effectiveness as a tyrosine kinase on other proteins and a decrease in self-phosphorylation. As yet no natural protein substrates other than the receptor itself have been identified.

12E The nature of the mitogenic signal

The effects of EGF and other growth factors on cells are very complex including both acute and long-term responses. There is an obvious similarity to the effects of insulin, but in the case of growth factors the long-term effects are more prominent and there seems to be an additional complication in that two signals acting together are necessary for a full response of the cell. It is clear that the effects of growth factors require at least one and possibly several different intracellular messengers. As with insulin, many different mechanisms have been proposed but there is no clear picture of the whole sequence of events.

12E1 Role of receptor cross-linking and internalization

The stimulation of DNA synthesis and internalization of EGF have similar dose–response curves. Both processes are half-maximally stimulated at 10^{-10} mol/l EGF which gives about ten per cent receptor occupancy. This suggests that the two processes may be linked and that the stimulation of DNA synthesis may be a consequence of the internalization process. It is worth noting that the mitogenic effect of EGF requires a rather high receptor occupancy compared with the occupancy needed for the acute effects of hormones described in the preceding chapters. Rapid internalization of hormone receptors usually requires a high degree of receptor occupancy.

There are three ways in which the cross-linking and internalization process could generate an intracellular signal.
1 The initial association of the receptors might lead to the formation of a second messenger inside the cell.
2 EGF itself, or a degradation product of EGF, might function as a second messenger after entry into the cell.
3 The EGF receptor, or a product derived from the receptor, might function as the second messenger.

There is evidence that the clustering process is important. As with insulin, EGF-receptor antibodies of the IgM type can mimic the effect of the hormone both in promoting receptor aggregation and in promoting the cellular response. An analogue of EGF can be made by cyanogen bromide treatment which cleaves the peptide chain next to the single methionine residue (see Fig. 12.1) yielding two chains linked by a disulphide bridge. This derivative binds to the EGF receptor but does not promote clustering. It is ineffective in promoting the mitogenic response, but the addition of antibodies to EGF restores the clustering of receptors and the mitogenic response. It seems likely, therefore, that receptor association is necessary for the physiological effects of EGF.

There is no evidence for the occurrence of intact EGF in the cytoplasm and if the hormone itself is to act as a source of intracellular messenger it would have to be a breakdown product. EGF degradation can be inhibited by a number of agents including chloroquine and methylamine which interfere with lysosomal function, colchicine, and inhibitors of cathepsin B such as leupeptin and antipain. Chloroquine and methylamine are highly toxic and cause a general impairment of cell functions. The other three inhibitors, however, are not toxic. They do not block the effect of EGF and, if anything, tend to potentiate the response. It seems likely, therefore, that EGF degradation is involved in down regulation and not in mediating the response.

The receptor itself seems to be degraded in a fairly coordinated way. Smaller polypeptides appear in the cell which are clearly products of receptor degradation, but there is no evidence that these fragments have any messenger function and this is unlikely since they are associated with the lysosomes rather than in the cytoplasm. This degradation is probably also concerned with down regulation.

There is good evidence that receptor aggregation is involved in the response. The possibility that the subsequent internalization step may also be necessary cannot be excluded since there have been no reports of experiments where internalization was prevented while allowing aggregation to take place. However, there is no evidence that any of the products of internalization have a messenger function.

12E2 Role of tyrosine kinase activity

The observation that the receptor has a tyrosine kinase activity is very interesting. Other examples of tyrosine kinases are the insulin receptor, the PDGF receptor and propbably other growth-factor receptors. The product of the $pp60^{v-src}$ oncogene of the tumour-promoting Rous sarcoma virus is a tyrosine kinase, and an apparently identical gene, pp60^{c-src}, occurs in normal mammalian cells. Several other tumour-promoting viruses also have genes which code for tyrosine kinases, and in the case of the oncogene v-erb-B the sequence of the protein product is very similar to the sequence of part of the EGF receptor. These viruses promote uncontrolled cell growth and division. An attractive hypothesis is that this results from excessive stimulation of a process normally controlled by natural tyrosine kinases as a result of the production of the tyrosine kinase coded for by the viral gene.

The circumstantial evidence that tyrosine-directed protein kinases are important in control of cell growth is very strong. However, nothing is known about the molecular mechanisms involved. The self-phosphorylation of growth-factor receptors may be an important step in mediating the effects of hormone binding, or phosphorylation of other proteins may be important. However, the function of proteins, other than the receptor, which are phosphorylated has not been determined. EGF cleaved with CNBr is effective in increasing receptor tyrosine kinase activity but is ineffective as a mitogen. Tyrosine kinase activation may therefore be necessary for the mitogenic effect but it is not sufficient.

12E3 Cytoplasmic protein factors

The cytoplasm of 3T3 cells contains factors which are able to stimulate DNA synthesis in cell-free nuclei. Extracts from quiescent cells in G_0/G_1 phase have little effect, but if the cells are incubated in the presence of EGF the cytoplasm acquires the ability to increase DNA synthesis. The EGF concentration dependence is similar to that for mitogenesis. Fractionation of the cytoplasm on a sucrose density gradient showed three peaks of activity with molecular weights of 46 000, 110 000 and 270 000. All three appear to be proteins since the activity is abolished by trypsin. Similar activating factors have been described in embryonic cells and in tumour cells.

As well as the factors appearing in the cytoplasm, EGF induces the secretion of several glycoproteins with a molecular weight of about 34 000. The role of these proteins is not known but their appearance appears to be related to the mitogenic response. Other growth factors also induce their secretion.

EGF also increases poly ADP-ribosylation in 3T3 cells. Chromatin proteins, both histones and non-histones, are poly ADP-ribosylated and the time course of the effect is consistent with a role in mitogenesis.

However, neither the mechanism by which poly ADP-ribosylation is induced by EGF nor the role, if any, of poly ADP-ribosylation in activating DNA synthesis has been characterized. There are also changes in DNA phosphorylation and DNA methylation in response to EGF, but again the physiological significance is unknown.

12E4 Cyclic nucleotides

There has been considerable interest in the role of cyclic nucleotides, in particular cyclic AMP, in the control of cell growth and proliferation. In 3T3 cells, a number of different agents which increase cyclic AMP concentration also stimulate mitogenesis. The effect is dependent upon the presence of one of several different mitogenic factors (Table 12.4), since elevating cyclic AMP does not appear to be mitogenic on its own. A number of different methods were used to increase the cellular cyclic AMP level including the addition of cholera toxin and inhibitors of phosphodiesterase such as isobutylmethylxanthine (IBMX). Both potentiated the effects of growth factors. The addition of analogues of cyclic AMP such as dibutyryl cyclic AMP also potentiates the response to growth factors.

The physiological role of cyclic AMP in the control of cell growth is unclear. Growth factors which affect 3T3 cells, such as EGF, FDGF, vasopressin and insulin, do not increase the cytoplasmic cyclic AMP concentration. Furthermore, it is possible to activate mitogenesis with combinations of effectors which do not increase cyclic AMP. It is not clear what the natural effector raising cyclic AMP levels is, but a possible candidate is adenosine. The adenosine analogue 5'N-ethylcarboxyamide-adenosine (NECA) is mitogenic in the presence of other growth factors and the effect is potentiated by inhibition of phosphodiesterase. NECA is selective for the activatory adenosine receptor (see Chapter 3). However, there is little evidence in other systems for a receptor-mediated stimulation of mitogenesis by adenosine. In many cell types, increasing cyclic AMP inhibits mitogenesis rather than activating it.

Table 12.4. Mitogens whose action is potentiated by elevated cyclic AMP in 3T3 cells

EGF
Vasopressin
FDGF
Insulin

12E5 Ion fluxes: Na$^+$ and Ca^{2+}

The addition of serum to 3T3 cells leads to a marked activation of the Na$^+$/K$^+$-ATPase which can be demonstrated by an increase in ^{86}Rb$^+$ influx. Rb$^+$ is transported in place of K$^+$ and acts as a K$^+$ analogue. The effect is inhibited by ouabain showing that the Na$^+$/K$^+$-ATPase is involved. Many individual growth-promoting agents will produce this effect including EGF, FDGF, PDGF, vasopressin and phorbol esters. If the Na$^+$/K$^+$ pump is inhibited by adding ouabain, growth factors increase the cellular content of Na$^+$ rather than of K$^+$. [^{22}Na]-Na$^+$ influx is increased as is the influx of Li$^+$, which acts as an analogue of Na$^+$. This indicates that the stimulation of the Na$^+$/K$^+$-ATPase is a consequence of increased influx of Na$^+$ ions rather than a direct activation of the pump. There is good evidence that increased cytoplasmic Na$^+$ does activate the Na$^+$/K$^+$-ATPase in intact cells, since, artificially increasing the plasma membrane Na$^+$ permeability with the ionophores monensin or gramicidin activates the ATPase.

The observation that many mitogens increase Na$^+$ influx raises two questions.
1 Does changing Na$^+$ influx by other means affect the mitogenic response?
2 If the change in Na$^+$ flux is important, how does this affect the mitogenic response?

Na$^+$ influx can be inhibited by reducing the Na$^+$ concentration in the medium, or by the addition of amiloride which inhibits a Na$^+$/H$^+$ exchange carrier in many cell types. Both treatments inhibit the mitogenic response.

Artificially increasing Na$^+$ influx activates DNA synthesis in the presence of growth factors. Monensin is not effective, probably because it increases the Na$^+$ permeability of the mitochondrial membrane as well as the plasma membrane, and thereby disrupts the energy metabolism of the cell. However, mellitin, a peptide found in bee venom, and the polyene antibiotic amphotericin B both specifically activate Na$^+$ fluxes across the plasma membrane and both stimulate mitogenesis in the presence of other growth factors. It seems that stimulation of Na$^+$ influx may be necessary for mitogenesis. It is not, however, sufficient since in all cases the presence of a growth factor is also required.

An increase in Na$^+$ influx might be expected to have several effects on the cell. If Na$^+$ influx is via an Na$^+$/H$^+$ exchange carrier in the plasma membrane, increased Na$^+$ influx might lead to alkalinization of the cytoplasm. Such a carrier is present in the plasma membrane of most cell types and can be identified by its sensitivity to inhibition by amiloride. The proliferation of many cell types in culture is very sensitive to changes in the extracellular pH, and there is some evidence for a slight alkalinization of the cytoplasm during the mitogenic response.

Increased cytoplasmic Na^+ might also lead to an increase in cytoplasmic Ca^+ either by stimulation of efflux from the mitochondria via a Na^+-dependent efflux carrier or via a plasma-membrane Na^+/Ca^{2+} exchange carrier (see Chapter 5). However, in most of the cell types used to study the mitogenic response, it is not established whether the mitochondrial Ca^+ efflux carrier is of the Na^+-dependent type or if there is a Na^+/Ca^{2+} exchange carrier in the plasma membrane.

There is a large body of evidence that Ca^{2+} may be involved in the mitogenic response. For example, mitogens often but not always change [$^{45}Ca^{2+}$] fluxes, and calcium depletion blocks mitogenesis. However, it is even more difficult to determine the significance of such experiments when the final response is 15–24 hours after the initial exposure to the agonist than when the response is rapid. A number of direct measurements have been made mostly using the Ca^{2+} indicator Quin 2. In lymphocytes, the lectin concanavalin A is mitogenic and causes a rapid increase in cytoplasmic free Ca^{2+}. In 3T3 cells, vasopressin and epidermal growth factor cause an immediate increase in cytoplasmic free Ca^+ concentration reaching a peak of about 0.5 µmol/l after 1–2 minutes. The concentration then declines to about 0.2 µmol/l, slightly above the resting level of 0.15 µmol/l. Other mitogens, including insulin and the phorbol ester, TPA, do not increase the cytoplasmic free Ca^{2+} level. It appears that an early increase in Ca^{2+} could be involved in the mechanism of action of some mitogens but not others. It also seems that increased influx of Na^+ is not necessarily caused by an increase in cytoplasmic free Ca^{2+} since TPA increases Na^+ influx but not cytoplasmic free Ca^{2+}.

12E6 The mitogenic signal: conclusions

There are many different proposals for the signal mediating the response to growth factors and other mitogens. In many cases it has been shown that the proposed signal is not sufficient on its own to activate mitosis and there is no case where a single effect can be shown to be common to all known mitogens. One of the most striking features of the activation of cell growth and proliferation is that a substantial response requires at least two mitogens. It seems likely that there are at least two independent mechanisms of activation and that both must be stimulated for an effective response.

12F Mechanism of action of NGF

As with other growth factors, NGF has rapid effects involving a general stimulation of metabolism and increased uptake of precursors for macromolecule synthesis. The long-term effects of NGF are somewhat different from other growth factors and this probably reflects the specialized function of neurons. There is a chemotactic response involved in the stimulation and direction of outgrowth of axons from the cell body. There is no effect on nerve-cell division in the adult animal but both RNA and protein synthesis are increased. Specific enzymes are induced, in particular tyrosine hydroxylase and dopamine β-hydroxylase, and the activity of these enzymes is often used as a convenient means of detecting a response to NGF. In a fully developed nerve cell most of the receptors are on the nerve terminal although there are some receptors on the cell body. The nuclei and ribosomes are in the cell body which in a large animal may be several feet from the terminal. Clearly this presents a problem of communication between the nerve terminal and the main body of the cell which does not arise in other cell types.

12F1 The NGF receptor

Two classes of NGF binding site can be identified with dissociation constants of 2×10^{-11} mol/l and 2×10^{-9} mol/l. Receptors on the surface of the cell body and the receptors on nerve terminals appear to be the same. Both high-affinity and low-affinity receptors seem to be physiologically significant. Binding is highly specific, quite minor chemical modifications of the β subunit of NGF causing a loss of binding to both types of receptor and a parallel loss of biological activity.

It has been suggested that high- and low-affinity receptors represent different populations of the same protein. A number of studies have been done with PC12 phaeochromocytoma cells. This is a rat cancer cell line derived from adrenal chromaffin cells. In the presence of NGF, cell division stops and neurite outgrowth can be observed. Thus the cells acquire some of the characteristics of normal nerve cells. On removal of NGF the neurites degenerate and cell division begins again. A major advantage of PC12 cells is that they can be grown in the absence of NGF. This makes it possible to examine the effects of adding NGF to a cell which has not been exposed to the hormone before. Normal nerve cells do not survive for prolonged periods in culture in the absence of NGF and therefore have always been exposed to NGF at some stage. An obvious disadvantage of the PC12 cell line is that they are not normal nerve cells.

PC12 cells have only low-affinity NGF receptors. After exposure to NGF for about 15 minutes, high-affinity receptors begin to appear. If cells are exposed to [^{125}I]-labelled NGF for a brief period the label appears first in the low-affinity receptors and then in the high-affinity receptors. Over a short period of time the total number of receptors does not change suggesting that low-affinity receptors are converted to high-affinity receptors (Fig. 12.6). Most of the biological responses to NGF appear to correlate with the occupancy of the high-affinity receptors.

12F2 Retrograde transport

Normal nerve cells have the ability to transport material from the cell body to the terminal (anterograde transport), or from the terminal to the cell body (retrograde transport). Retrograde transport of NGF was first demonstrated by injection of [^{125}I]-labelled NGF into the iris of mice. The label accumulated in the dorsal root ganglia. The accumulation was specific as label only accumulated in the ganglia associated with the innervation of the iris. Furthermore, if the labelled NGF was injected in the iris on only one side of the animal, label only accumulated in the ganglia on the same side and not on the other side. The accumulation was receptor specific since a number of other proteins were tested and were not taken up. On the other hand, other proteins that were known to have receptor sites on the nerve terminal, including tetanus toxin, cholera toxin, and a number of lectins, were taken up and accumulated in the ganglia. A reduction in the number of NGF receptors on the nerve terminal could be detected. In PC12 cells there is evidence that the high-

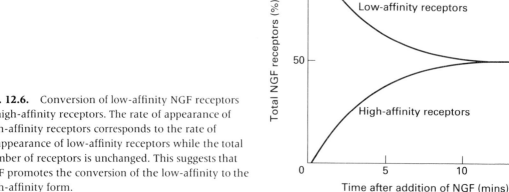

Fig. 12.6. Conversion of low-affinity NGF receptors to high-affinity receptors. The rate of appearance of high-affinity receptors corresponds to the rate of disappearance of low-affinity receptors while the total number of receptors is unchanged. This suggests that NGF promotes the conversion of the low-affinity to the high-affinity form.

affinity receptors are internalized and replenished on the cell surface from the low-affinity receptor pool.

Receptor-mediated internalization of ligands is a widespread phenomenon which seems also to occur at nerve terminals. The subsequent transport of the internalized ligand down the axon to the cell body is of necessity unique to nerve cells. NGF is transported at a rate of between 5 and 10 mm/min as ligand–receptor complexes inside vesicles which are derived from the plasma membrane. The process requires energy and is inhibited by disruption of microtubles by colchicine or vinblastine. During transport NGF remains intact. On arrival in the cell body, the transport vesicles fuse with lysosomes and NGF is degraded (Fig. 12.7).

12F3 Physiological importance of retrograde transport

Receptor-mediated uptake and retrograde transport might be important in the mechanism of action of NGF. Alternatively, it could simply be a down-regulation mechanism as is observed with many hormones. In the case of NGF, the completion of a down-regulation process by degradation of the hormone, and possibly also the receptor, would require retrograde transport since lysosomes which are the site of breakdown only occur in the cell body and not in the nerve terminals. It is important to distinguish between these two possibilities since, if retrograde transport is concerned with down regulation, there must be another second messenger to mediate the cellular response. Retrograde transport can be prevented from interfering with microtubule function, with agents such as vinblastine, by physical damage to the axon between the terminal and the cell body, and by chemical damage to the nerve terminal using 6-hydroxy dopamine. All of these treatments cause a rapid reduction in the activity of tyrosine hydrolylase and dopamine β-hydroxylase, and degeneration of the ganglia. These effects can be reversed by supplying NGF directly to the ganglia where it will interact with receptors on the cell body. It seems likely, therefore, that the retrograde transport of NGF bound to its receptor is required to generate a signal in the cell body and produce a response.

12F4 Nature of the signal

The first possibility to consider is that NGF itself functions as a messenger in the cell body, but there is no evidence that free NGF is released into either the cytoplasm or the nucleus. If NGF does act as a second messenger, then increasing the concentration of NGF in the cytoplasm should cause a response. With a large peptide like NGF this is

204 *Chapter 12*

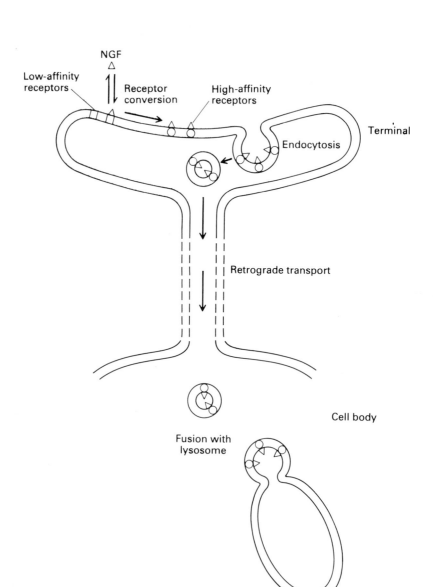

Fig. 12.7. Uptake and retrograde transport of NGF.

difficult, but it has been achieved by preparing erythrocyte ghosts containing NGF and fusing them with the cell body of nerve cells. This gave substantial concentrations of NGF in the cytoplasm but did not activate the cellular response. The inserted NGF was degraded at a much slower rate than NGF arriving in the cell body by retrograde transport suggesting that its location is different. An obvious problem with this type of approach is that the fusion process might damage the normal functioning of the cell and that this might be the reason for a lack of response. However, cells containing inserted NGF responded normally to NGF supplied to the nerve terminals showing that the normal response was intact. In common with other growth factors it seems very unlikely that NGF itself functions as an intracellular messenger.

12F5 Cyclic nucleotides and inorganic ions

There is no evidence that NGF affects the activity of adenylate cyclase or phosphodiesterase, or that it affects the concentration of cyclic AMP in the cell. Alterations in the concentration of cyclic AMP in nerve cells do not affect cell growth.

There has been relatively little attention paid to a possible role for Na^+ or Ca^{2+} in the action of NGF. However, NGF does activate the Na^+/K^+-ATPase and this implies a change in the Na^+ gradient. NGF also increases phosphatidylinositol turnover, and in all other systems which have been studied this is thought to be associated with a rise in cytoplasmic free Ca^{2+}. While there is little evidence for a role of inorganic ions in the action of NGF this possibility might repay investigation.

12F6 Nuclear accumulation of NGF

There have been a number of reports of NGF receptors in the nucleus of nerve cells and also that as much as thirty per cent of [^{125}I]-labelled NGF is found in the nucleus after retrograde transport. This conflicts with other observations in which association of [^{125}I]-labelled NGF with the nucleus cannot be detected by autoradiography.

PC12 cells appear to take up NGF as a hormone−receptor complex in the same way as normal nerve cells and at first the NGF is degraded in the lysosomes. However, after about 5 hours [^{125}I]-NGF can be detected in the nucleus. Blocking lysosomal function wiith chloroquine removes the lag in the appearance of NGF in the nucleus and increases the extent of accumulation (Fig. 12.8) suggesting that lysosomal degradation and nuclear accumulation may be competing processes (Fig. 12.9). The accumulation of NGF in the nucleus takes 5−6 days to complete. The hormone is accumulated as a hormone−receptor complex and is associated with the nuclear membrane rather than the chromatin. NGF extracted from the nucleus is intact and biologically active: the nuclear receptors have very similar properties to the high-affinity receptors in the cell membrane, and their number increases as a result of treatment of the cell with NGF. This observation has led to the suggestion that NGF−receptor complexes are inserted into the nuclear membrane by

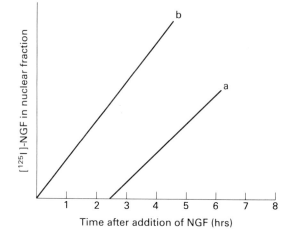

Fig. 12.8. Appearance of NGF in the nucleus. The experiment used PC12 cells: (a) shows the rate of appearance of NGF in the nucleus of control cells, and (b) the rate in cells where the lysosomal function had been blocked by addition of chloroquine.

fusion of receptor-containing vesicles after retrograde transport. If the nuclear receptors have a physiological role this would explain why free NGF in the cell body is ineffective in producing a response.

These proposals raise a number of questions.

1 Why is nuclear accumulation not detected by autoradiography of normal nerve cells after exposure to [^{125}I]-NGF? The initial phase of lysosomal degradation may account for this. In PC12 cells almost all the internalized NGF is degraded for the first five hours. In normal cells this period might be even longer. Detection of nuclear binding would therefore require a very long exposure to [^{125}I]-NGF.

2 The lag in accumulation appears to be caused by competition from lysosomal degradation. It can be abolished by inhibitors of lysosomal function or by cytochalasin which disrupts actin filaments. This suggests that the direction of receptor-containing vesicles to the lysosomes requires the intervention of contractile filaments while the accumulation in the nuclear membrane does not. There is, however, no satisfactory explanation for the switch from fusion of receptor-containing vesicles with the lysosomes to fusion with the nuclear membrane.

3 The most important question is whether or not the nuclear accumulation is important in the biological response. In PC12 cells the accumulation is detectable before any outgrowth of neurites can be detected, and the number of cells which form neurites correlates well with the extent of nuclear accumulation. After removal of NGF from the medium, the nuclear content of NGF declines before the degeneration of the neurites and the re-initiation of cell division.

4 If it is accepted that the nuclear NGF–receptor complexes are

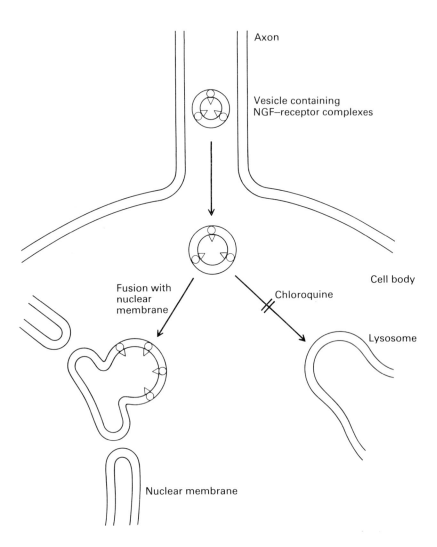

Fig. 12.9. Postulated model for accumulation of NGF in the nucleus.

important in the mechanism of action of the hormone, an additional second messenger will still be needed since the NGF–NGF receptor complexes are located in the nuclear membrane and the response requires that chromatin function be changed.

The observation of nuclear receptor–NGF complex accumulation is very interesting and at least a preliminary case can be made for its function and significance. However, almost all the data have been obtained with an abnormal cell line and the conclusions should perhaps be treated with caution until a similar course of events can be demonstrated in normal cells.

12G Conclusions

Cell growth and proliferation are long-term complex processes. It is not surprising, therefore, that their regulation is complex and requires multiple factors. Many individual responses have now been identified and in some cases a connection with the mechanism of action of carcinogens is apparent. As yet the relationships of the multiple responses to each other, and their relative importance in the overall responses of the cell, are unclear.

Suggestions for further reading

Carpenter G. & Cohen S. (1979) Epidermal growth factor. *Annu Rev Biochem,* **48**, 192–216.

Das M. (1983) Epidermal growth factor receptors and mechanisms of mammalian cell division. In *Current Topics in Membranes and Transport* (Ed. by A. Kleinzeller and B.R. Martin), vol. 18, pp. 381–406. Academic Press.

Rozengurt E. (1981) Stimulation of sodium influx Na/K pump activity and DNA synthesis in quiescent cultured cells. *Advances In Enzyme Regulation,* **19**, 61–85.

Rozengurt E. (1983) Growth factors, cell proliferation and cancer, an overview. *Mol Biol Med,* **1**, 169–182.

Shooter E.M., Yankner B.A., Landreth G.L. & Sutter A. (1981) Biosynthesis and mechanism of action of nerve growth factor. *Recent Progress In Hormone Research,* **3**, 417–447.

Thoenen H. & Barde Y.A. (1980) Physiology of nerve growth factor. *Phsyiol Rev,* **80**, 1284–1335.

Chapter 13 Metabolic integration: general introduction

13A Introduction

The previous chapters have been concerned with the mechanisms by which hormones produce changes in cell function. The rest of the book will describe the regulation of intermediary metabolism in muscle, liver and adipose tissue, and the interactions between these tissues through the circulation. The metabolic pathways in the three tissues are well described and the regulation is well characterized. In each case the major physiological role of the tissue is different, and it is possible to find examples of many types of interaction between different regulatory mechanisms. The aim of this section of the book is not to provide a comprehensive description of the regulation of metabolism at the level of the whole cell, but rather to show how different types of control operate at different levels to produce a concerted response to a change in the metabolic state of the whole animal. A number of general points can be listed.

1 Effects on one tissue may lead to effects on another tissue as a result of the release of metabolites into the circulation.
2 Several hormones may interact to produce a particular effect at the level of the whole cell.
3 A single hormone may act on a cell using more than one mechanism.
4 Effects of hormones and other extrinsic regulators take place against the background of the intrinsic control mechanisms of the cell. Intrinsic and extrinsic mechanisms can operate on the same processes and the metabolic state of the cell can affect the nature of the response to hormones.
5 The rate of flux in a particular pathway may increase in response to a hormone in one tissue, while in a different tissue the same hormone may inhibit the pathway. This reflects the different physiological roles of different tissues and can be explained by the types of interaction described above under 2, 3 and 4.

In the rest of this chapter, the physiological functions of muscle, liver and adipose tissue, and the metabolic relationships between the three tissues and the brain, will be described.

13B Physiological functions of muscle, liver and adipose tissue

13B1 Muscle

The function of muscle is mechanical and its main regulatory requirement is the maintenance of adequate supplies of ATP for contraction. In heart and skeletal muscle, metabolism is largely oxidative and either glucose, fatty acids or ketone bodies can be used as fuel. An important control function, therefore, is to determine which fuel source is used so as to take account of the needs of the whole organism. In particular, when the availability of glucose in the diet is low, glucose utilization is inhibited and the muscle derives most of its energy from lipid sources. A second important regulatory function is the control of the build up of energy resources as glycogen, and to a lesser extent triacylglycerol, in the muscle during relaxation. Finally, during prolonged starvation, a proportion of muscle protein is broken down to provide precursors for the maintenance of blood glucose.

13B2 Liver

Liver metabolism is very complex. It is the organ which is first exposed to blood coming from the gut via the hepatic portal vein, and a major function is to respond to the composition of the nutrients available in the diet, and to make it possible for the organism as a whole to make the most efficient use of what is available. Liver is able to perform virtually all the interconversions between lipid, carbohydrate and amino acids, which are possible in a mammalian cell. It is the only organ able to produce ketone bodies from fatty acids, and the only site of urea biosynthesis, which is the means of eliminating excess nitrogen. It is a major site for fatty-acid and steroid biosynthesis, and the only site for the processing of dietary lipid into lipoproteins. The liver also stores glucose as glycogen and synthesizes glucose via gluconeogenesis, a critical function during starvation.

In the well-fed state, the liver first builds up its glycogen reserves and then switches to fatty acid biosynthesis. The conventional view is that circulating glucose provides the main substrate for both of these processes. However, recent work suggests that, while increased blood glucose is important as a signal for increased fatty-acid and glycogen synthesis, the main sources of carbon are lactate and fructose. Fructose will be particularly important in the West, where sucrose is a major component of the diet. In the fed state, another important role of the liver is to process dietary lipid from chylomicrons to lipoproteins.

During starvation, the liver releases glucose, first from stored glycogen and then as the end-product of gluconeogenesis. In prolonged starvation the main source of carbon for gluconeogenesis is amino acids, and this results in an increase in the activity of the urea cycle in order to dispose of the nitrogen content. Under these conditions the liver uses fatty acids for its own energy needs and also converts fatty acids to ketone bodies. After a relatively long period of starvation, ketone bodies supply most of the energy requirements of the heart, a substantial part of the needs of skeletal muscle, and about seventy-five per cent of the needs of the brain.

13B3 Adipose tissue

The metabolism of adipose tissue is relatively simple. In many animals, including the rat, it converts glucose to fatty acids and triacylglycerol and also stores lipid derived from the diet and from synthesis in the liver. In other species, including humans, the storage function is by far the more significant and there is relatively little synthesis from glucose. During fasting or exercise, fat synthesis and storage is swtiched off and the stored lipid is mobilized as fuel for other tissues. This comparatively simple situation, and the fact that rat fat cells respond to many different hormones, has made it a particularly attractive tissue in which to study mechanisms of hormone action.

13C Absorption and transport of metabolites

13C1 Carbohydrate

Carbohydrate in the diet is digested to monosaccharides and absorbed in the small intestine. There are three main sugars: glucose, galactose and fructose. The intestinal lumen membrane has two carriers, one to transport glucose and galactose and one to transport fructose. Both are able to transport the sugars against a concentration gradient and therefore require a source of energy. In both cases this is supplied by the Na^+ gradient which is maintained by the Na^+/K^+-ATPase. Na^+ ions are co-transported together with the sugar into the cell down the Na^+ gradient across the cell membrane (Fig. 13.1).

Carbohydrate is also transported in the circulation as glucose, galactose or fructose. The level of glucose is closely regulated and glucose may be derived from the diet or released from glycogen stores in liver, or synthesized and released by liver using the gluconeogenic pathway. Galactose and fructose only come from the diet. In most tissues, galactose probably enters the cell on the glucose transporter, which is regulated by insulin in muscle and adipose tissue, but not in liver, which is probably the site of most galactose metabolism. Fructose is taken up from the circulation by a different carrier which is not affected by

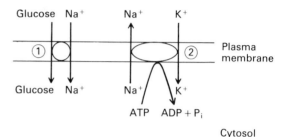

Fig. 13.1. Na$^+$-dependent glucose transport. (1) Co-transport of glucose and Na$^+$ into the cell. Energy is provided by the concentration gradient of Na$^+$, the intracellular concentration being about 10% of the extracellular concentration. (2) The Na$^+$/K$^+$-ATPase which maintains the Na$^+$ gradient at the expense of ATP.

insulin. In adipose tissue and muscle, it is possible to detect small effects of insulin on fructose uptake *in vitro* when no glucose is present. This probably results from a small amount of transport on the glucose carrier being activated by insulin. Glucose is always present outside the cell *in vivo* and will compete against fructose for its carrier so that under natural conditions fructose transport on the glucose carrier is unlikely. Both galactose and fructose can be metabolized directly in most tissues but the liver is the major site. Galactose is metabolized by the pathway shown in Fig. 13.2. It generates glucose 1-phosphate so its metabolism is subject to most of the controls which operate upon glucose metabolism (see Chapters 14, 15, and 16). In the liver, fructose is metabolized by the pathway shown in Fig. 13.3, involving a specific fructose kinase which generates fructose 1-phosphate. This bypasses phophofructokinase, the main site of regulation in glycolysis. In adipose tissue or muscle, fructose is converted to fructose 6-phosphate by hexokinase and enters glycolysis. In quantitative terms, however, this is a minor pathway compared to the metabolism of fructose in liver.

13C2 Lipid

The main lipid component in the diet is long-chain triacylglycerol. This is digested in the gut to form monoacylglycerol and free fatty acids. These are taken up in the small intestine and reconverted to triacylglycerol which is combined with phospholipid, cholesterol, cholesterol esters and a specific protein, apolipoprotein B, to form chylomicrons. Chylomicrons are released into the lymphatic system by exocytosis, and from there enter the circulation. Chylomicrons have a diameter of about 70 nm and consist of about 85% triacylglycerol, 8% phospholipid, 2% cholesterol, 3% cholesterol ester and 2% protein. In the circulation they

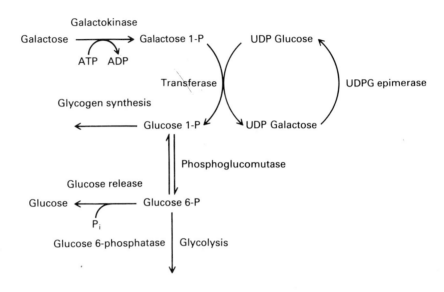

Fig. 13.2. Reactions of galactose metabolism in liver.

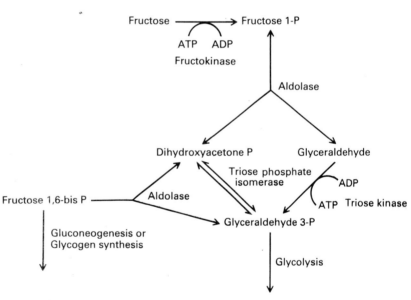

Fig. 13.3. Fructose metabolism in the liver. At the point where dihydroxyacetone phosphate and glyceraldehyde 3-phosphate are formed, fructose metabolism coincides with glucose metabolism. Carbon flow may then be in the glycolytic or gluconeogenic direction.

acquire an additional protein component, apolipoprotein C, which renders them susceptible to attack by lipoprotein lipase. This hydrolyses triacylglycerol to fatty acids and glycerol, depleting the chylomicron of its lipid. Fatty acids are taken up by a number of tissues using this mechanism. Most are probably taken up by adipose tissue for storage but some will be taken up by the liver. Heart and skeletal muscle also contain lipoprotein lipase. In most tissues, lipoprotein lipase is attached to the surface of the capillaries, but in liver it is on the surface of the liver cell itself. The gradual removal of lipid from chylomicrons leaves a remnant particle which is taken up and metabolized by the liver.

The liver is also the site of synthesis of very low-density lipoproteins (VLDL). These are synthesized from triacylglycerol, made in the liver from fatty acids synthesized from glucose or from fatty acids taken up from the circulation. They are released into the circulation where they act as a source of fatty acids for adipose tissue. A little may be taken up by muscle.

13D Interactions between brain, muscle, liver and adipose tissue through the circulation

The balance between uptake and release of metabolites depends upon the availability in the diet of carbohydrate, lipid and protein which will appear in the circulation as glucose, fatty acid and triacylglycerol, and amino acids. The balance between the three types of food will vary

Fig. 13.4. (Opposite) Metabolic fluxes in different dietary conditions. This shows the main direction of carbon flow between brain, adipose tissue, liver and skeletal muscle in (1) the well-fed state (bold line), (2) the fasted/starved state (fine line) and (3) during muscle exercise (broken line). All the available pathways are shown, but it is not possible to give an indication of their relative rates in the diagram.
(a) Brain. In the fed state, glucose uptake provides almost all the energy needs of the brain. During starvation, there is a progressive increase in the use of ketone bodies which eventually replace 75% of the glucose requirement. (b) Adipose tissue. In rats in the fed state, about 50% of stored triacylglycerol is derived from circulating lipoprotein and 50% from the new synthesis from glucose. In humans, almost all stored triacylglycerol is derived from circulating lipoprotein. In the starved state, metabolism switches to lipolysis and release of fatty acids and glycerol. (c) Muscle. Under certain conditions, the major source of carbon for energy is circulating free fatty acids. Muscle can metabolize triacylglycerol as circulating lipoprotein and may store a little triacylglycerol as well. However, the contributions of those pathways are very small relative to the energy demands of the cell. In starvation, the preference for fatty acids, as opposed to glucose, is intensified, and ketone bodies are also oxidized, particularly in the heart. If the muscle becomes anaerobic, glycolysis becomes the energy source using glucose from the circulation and from stored glycogen. (d) Liver. Liver utilizes no glucose in the starved state, and little glucose in the fed state. In the fed state, dietary triacylglycerol as circulating lipoprotein is processed and mostly re-exported into the circulation. There is also some synthesis *de novo* of fatty acids and triacylglycerol, most of which is also exported. The main carbon sources in the fed state are lactate, amino acids from the diet, and sugars other than glucose—mainly fructose and galactose. These are used for energy, for fatty acid synthesis, and for glycogen synthesis. In starvation, these carbon sources are used for gluconeogenesis and energy from fatty oxidation. Fatty acids are also used for ketone body biosynthesis.

Metabolic integration: general introduction

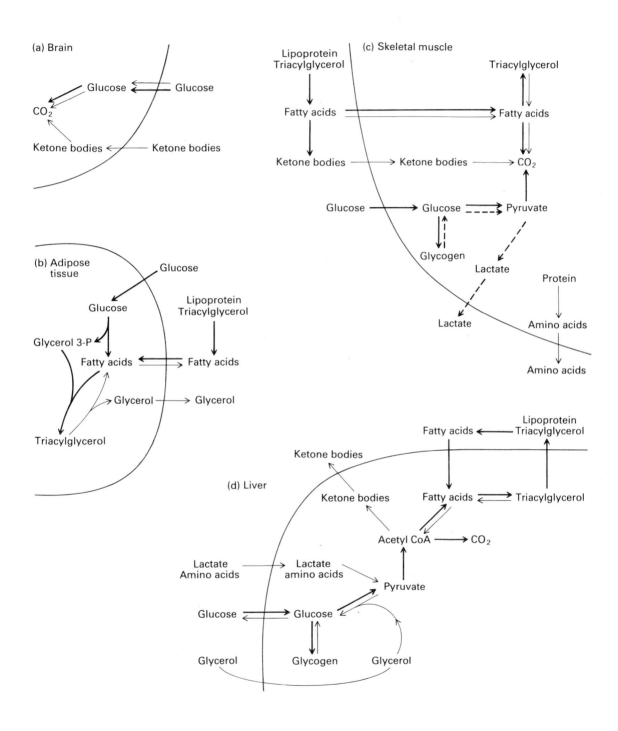

depending upon the type of diet. Perhaps the easiest way to describe the interaction between the tissues is to compare two extreme dietary states. In the first there is an abundance of carbohydrate and hence of glucose: in the second, starvation, there is no food intake at all.

The flow of metabolites between the different tissues and the circulation in the two different states is shown in Fig. 13.4. In both cases the brain is consuming glucose since it has an absolute requirement for glucose as an energy source. During prolonged starvation, about 75% of this requirement can be supplied by ketone bodies as an alternative, and glucose uptake by the brain is reduced as ketone body levels in the circulation rise.

Heart and skeletal muscle (red muscle) tend to oxidize fatty acids rather than glucose under most conditions. The preference is intensified by starvation and ketone bodies are also oxidized, particularly in the heart. In the case of skeletal muscle the question of whether or not the muscle is at rest or exercising is also relevant. At rest when there is glucose available in the diet, muscle will build up its glycogen stores to be used as a rapidly available energy reserve during exercise. Unlike fatty acids, glucose can yield energy under anaerobic conditions, producing lactate as the end-product of glycolysis. Under violent exercise the rate of entry of oxygen into muscle can become rate limiting so that glycolysis becomes the only source of ATP. White muscle has a relatively poor oxidative capacity so that glycolysis is always important.

When dietary glucose is abundant the liver will take up both glucose and lipid derived from the diet. Glucose acts as the main source of energy and also as substrate for fatty acid and triacylglycerol synthesis. Dietary lipid is also used for triacylglycerol biosynthesis. Glucose will be stored as glycogen and there is some storage of triacylglycerol but most is released into the circulation as lipoprotein complexes. In starvation, liver depends upon fatty acid oxidation for energy, and fatty acids are also converted to ketone bodies which are released into the circulation as an additional energy source for red muscle and brain. Stored glycogen is broken down to release glucose into the circulation during the early stages of starvation, and gluconeogenesis from lactate, pyruvate, glycerol derived from triacylglycerol, and amino acids begins. As starvation continues, glycogen resources become depleted and gluconeogenesis becomes more important. Net protein breakdown in other tissues begins to provide gluconeogenic substrates. Finally, the liver begins to make use of its own protein for glucose synthesis.

In the well-fed state, adipose tissue will convert glucose to triacylglycerol which is stored. Dietary lipid and lipid derived from the liver will be taken in as fatty acid as a result of the action of lipoprotein lipase on triacylglycerols in circulating lipoproteins. This will also be converted to triacylglycerol. During starvation triacylglycerol will be broken down and released into the circulation as free fatty acids and glycerol. The fatty acids will be used for energy by muscle and liver and as a source of ketone bodies. The glycerol will be used as substrate for gluconeogenesis.

In the following chapters of the control of muscle, liver and adipose tissue metabolism will be considered in detail.

Suggestions for further reading

Newsholme E.A. & Leach A.R. (1983) *Biochemistry for the medical sciences.* Wiley. This describes the integration of metabolism very completely.

Chapter 14 Regulation of metabolism in cardiac and skeletal muscle

14A Introduction

The major pathways of intermediary metabolism in cardiac muscle are summarized in Fig. 14.1 The main regulatory requirement of muscle is to maintain the availability of ATP for contraction. As a result the balance of metabolism is essentially catabolic, synthetic pathways being confined to the laying down of fuel reserves and the maintenance of muscle structure. Muscle uses two main fuels: glucose which can be stored as

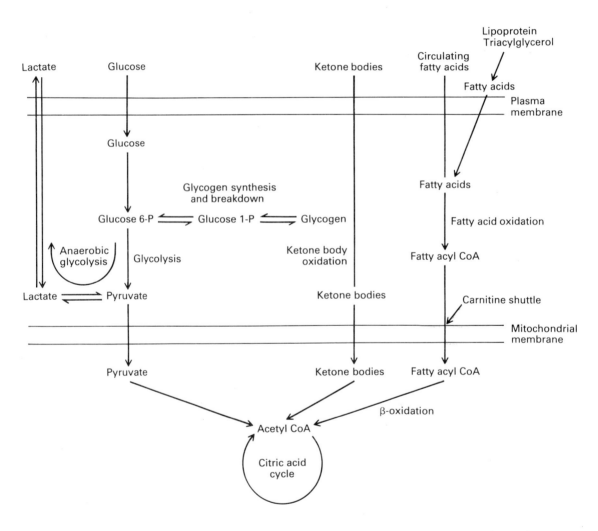

Fig. 14.1. Energy metabolism in muscle. This shows the major pathways for energy production in skeletal muscle. With the exception of glycolysis, all the pathways are dependent on oxygen.

glycogen, and lipid, in the form of fatty acids, which may be stored as triacylglycerol or ketone bodies. Under anaerobic conditions only glucose may be used since anaerobic glycolysis yields ATP while ATP synthesis from lipid breakdown is completely dependent upon oxygen. When there is a plentiful supply of oxygen red muscle tends to oxidize lipid substrates in preference to glucose.

14B Glycolysis and glycogen metabolism

The pathways of glycolysis and glycogen metabolism are shown in Fig. 14.2. The two pathways come together at glucose 6-phosphate and two steps are common to both processes: the entry of glucose into the cell and phosphorylation of glucose to glucose 6-phosphate by hexokinase. This part of the pathway must operate under conditions of both high and low energy demand. It is needed to fuel glycolysis under conditions of high energy demand and to supply the precursors for glycogen synthesis under conditions of low energy demand. A second branch point occurs at dihydroxyacetone phosphate which may be converted to either glyceraldehyde 3-phosphate, in which case it continues down glycolysis, or to glycerol 3-phosphate which can be used together with fatty acids to form triacylglycerol. In muscle, triacylglycerol stores are small in relation to the energy demands of the cell. A third branch point occurs at pyruvate which may either be converted to lactate, allowing glycolysis to function anaerobically, or enter the mitochondria to be converted to acetyl CoA and enter the citric acid cycle.

14C Regulation of muscle glycolysis

The function of glycolysis in muscle is to provide ATP for contraction, both directly and as a source of acetyl CoA to fuel the citric acid cycle. An increase in ATP demand should lead to an increase in glycolytic flux, and a likely signal for the activation of glycolysis is a fall in the concentration of ATP. The regulation of glycolysis has been studied extensively in the isolated perfused rat heart. Making the heart anaerobic increases glycolytic flux by a factor of 9 while the ATP concentration falls by only 20%. In insect flight muscle similar decreases in ATP level may be associated with a one hundredfold increase in glycolytic flux. This sugests that there must be a mechanism for the amplification of the initial signal, a fall in ATP concentration, to cause a relatively much larger increase in glycolytic flux. This is physiologically necessary since the purpose of the regulation is to maintain ATP levels.

There are three control points in muscle glycolysis: phosphofructo-

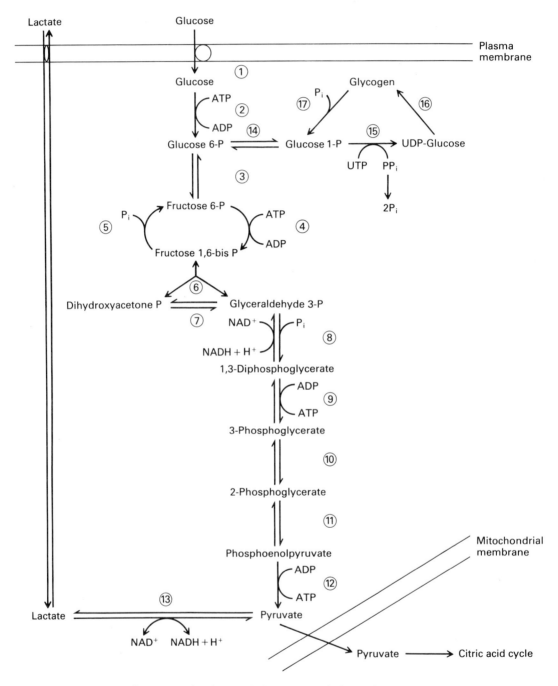

Fig. 14.2. Glycolysis and glycogen metabolism. The reactions involved are: (1) glucose transport, (2) hexokinase, (3) glucose 6-P isomerase, (4) phosphofructokinase, (5) fructose 1, 6-bisphosphatase, (6) aldolase, (7) triose

phosphate isomerase, (8) glyceraldehyde 3-P dehydrogenase, (9) phosphoglycerate kinase, (10) phosphoglyceromutase, (11) enolase, (12) pyruvate kinase, (13) lactate dehydrogenase, (14) phosphoglucomutase, (15) glucose 1-P uridyl transferase, (16) glycogen synthetase, (17) phosphorylase.

kinase, hexokinase and the transport of glucose into the cell. Their identification was described in Chapter 1.

14C1 Phosphofructokinase

Phosphofructokinase is the main control site in glycolysis. It catalyses the reaction:

fructose 6-phosphate + ATP $\longrightarrow$ fructose 1,6-bisphosphate + ADP.

As the end-product of glycolysis, ATP might be expected to act as a feedback inhibitor as well as a substrate. This can be tested by varying the concentration of ATP and measuring the rate of reaction. If this is done it is found that as the ATP concentration is increased the reaction rate increases, passes through a maximum at 1–2 mmol/l ATP and then declines (Fig. 14.3). This implies that in addition to the catalytic site, phosphofructokinase has a second binding site for ATP which has a regulatory function leading to inhibition of the enzyme. The ATP concentration in resting muscle is about 5 mmol/l and at this concentration the enzyme is substantially inhibited. A fall in ATP of 20% may be associated with a tenfold increase in glycolytic flux but would not on its own increase phosphofructokinase activity by more than 50% or so. Clearly additional mechanisms must operate and the inhibition by ATP is modulated by a number of other factors.

Inhibition is opposed by physiological concentrations of AMP. All tissues contain an enzyme adenylate kinase which catalyses the reaction:

ATP + AMP $\rightleftharpoons$ 2 ADP.

The equilibrium constant $K = [ADP]^2/[ATP][AMP]$

and is close to 1. The reaction is also close to equilibrium in the cell so that as the concentration of ATP falls, the concentration of both ADP and AMP will rise. In resting muscle the concentration of AMP is about 2% of the concentration of ATP. This has the effect that a small fall in the ATP level causes a relatively large rise in the concentration of AMP. For example, a 15% decrease in ATP results in a threefold rise in the concentration of AMP (see Table 14.1). The rise in AMP activates phosphofructokinase by reversing the inhibition by ATP. This has the

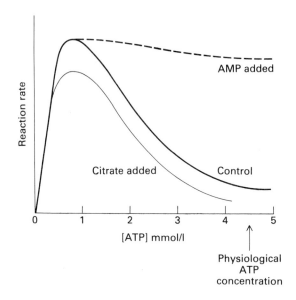

Fig. 14.3. Inhibition of phosphofructokinase by ATP.

Table 14.1. Effects of anoxia on adenine nucleotide concentrations in muscle

	Concentration (mM)		% Change
	Aerobic muscle	Anaerobic muscle	
ATP	10.0	8.5	−15
ADP	2.0	3.0	+50
AMP	0.2	0.8	+300

effect of amplifying the original signal which is the fall in ATP concentration.

Inorganic phosphate also relieves the inhibition of phosphofructokinase by ATP and this has the effect of further amplifying the response to a fall in ATP concentration. If the hydrolysis of ATP to ADP and P_i was the only source of inorganic phosphate, a small fall in the level of ATP would not be expected to change the concentration of P_i very much. However, muscle contains creatine phosphate which helps to maintain the concentration of ATP by the reaction:

creatine phosphate + ADP $\rightleftharpoons$ creatine + ATP.

Hydrolysis of ATP results in the overall reaction:

creatine phosphate $\longrightarrow$ creatine + P_i.

As a result, the P_i concentration may rise by substantially more than the fall in the concentration of ATP.

Inhibition by ATP is also reversed by the other substrate, fructose 6-phosphate, and by fructose 1,6-bisphosphate, one of the products. Activation of a rate-limiting enzyme leads to an increase in the concentration of the products (see Section 1F2). If the product, in this case fructose 1,6-bisphosphate, is itself an activator, this will have the effect of further increasing the response to any other activator of the enzyme. Product activation also increases the sensitivity to inhibition since inhibition of a rate-limiting enzyme leads to a fall in the product concentration. It seems likely that the role of fructose 1,6-bisphosphate activation is to increase the sensitivity of phosphofructokinase to both activation and inhibition by other effectors. The physiological role of activation by fructose 6-phosphate is unclear. Activation of phosphofructokinase will be associated with a fall in the concentration of its substrates, one of which is fructose 6-phosphate. In fact the changes in fructose 6-phosphate concentration, associated with changes in glycolytic flux, are quite small.

The inhibitory effect of ATP is intensified by citrate. This acts as a signal of the availability of alternative sources of ATP such as fatty acids or ketone bodies. In heart, metabolism of either lipid-derived substrate increases the concentration of citrate—the reasons will be considered later (see Section 14J). This control will contribute to an increase in rate in anaerobic conditions since lipid cannot be metabolized in the absence of oxygen. In starvation, this control will be important as a means of preventing glucose oxidation in favour of lipid oxidation and thereby sparing glucose for the brain. The ATP/AMP ratio will be a significant factor in controlling the activity of phosphofructokinase in muscle, under all conditions.

14C2 The phosphofructokinase/fructose bisphosphatase substrate cycle

Phosphofructokinase is activated by at least four different effectors, each by reducing the ATP inhibition. AMP, fructose 1,6-bisphosphate and inorganic phosphate all tend to amplify the response to a reduction in ATP concentration. However, the extent of the possible activation is limited by the increase in activity from maximal ATP inhibition to no ATP inhibition (Fig. 14.3). A larger response requires a different mechanism. Muscle has no obvious need to reverse the carbon flow in glycolysis. Nevertheless, the enzyme fructose bisphosphatase, which catalyses the conversion of fructose 1,6-bisphosphate to fructose 6-

phosphate, is present. This allows a substrate cycle between fructose 6-phosphate and fructose 1,6-bisphosphate (Fig. 14.4). The role of this substrate cycle is to amplify the activation of glycolysis resulting from a fall in ATP and a rise in AMP (see Section 1F4). The net flux through glycolysis will be the rate through phosphofructokinase minus the rate through fructose bisphosphatase. AMP activates phosphofructokinase and inhibits fructose bisphosphatase. As a result, glycolytic flux increases by more than the activation of phosphofructokinase. The greater the rate of flux through fructose bisphosphatase, and accordingly the rate of substrate cycling, the greater the extent of amplification. In insect flight muscle the rate of phosphofructokinase/fructose bisphosphatase substrate cycle is high and glycolytic flux can be activated by as much as one hundredfold.

14C3 Hexokinase

The regulation of hexokinase is comparatively simple: it is inhibited by its product glucose 6-phosphate. In muscle, glucose 6-phosphate is the substrate for two reactions: phosphoglucomutase which converts it to glucose 1-phosphate, and glucose 6-phosphate isomerase which converts it to fructose 6-phosphate (Fig. 14.5). Both these reactions are close to equilibrium in the cell, so that a rise or fall in the concentration of either fructose 6-phosphate or glucose 1-phosphate will result in a corresponding change in the concentration of glucose 6-phosphate.

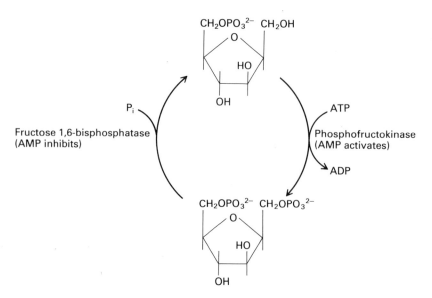

Fig. 14.4. The fructose 6-phosphate/fructose 1,6-bisphosphate substrate cycle.

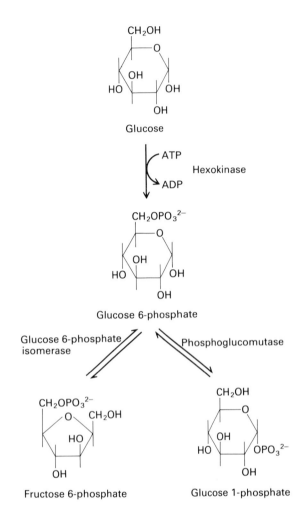

Fig. 14.5. Interconversion of hexose monophosphates. The phosphoglucomutase and glucose 6-phosphate isomerase reactions are close to equilibrium so that a change in the concentration of one hexose monophosphate leads to a corresponding change in both the others.

The control by glucose 6-phosphate is important in two different situations. Inhibition of phosphofructokinase will result in a rise in fructose 6-phosphate and this will lead to a rise in glucose 6-phosphate. If hexokinase was not inhibited by the rise in glucose 6-phosphate the concentration of both glucose 6-phosphate and fructose 6-phosphate would continue to increase. Since fructose 6-phosphate is an activator of phosphofructokinase, this might eventually override the inhibition of the enzyme with a resulting loss of control. The regulation of hexokinase is also important in maintaining the supply of glucose 1-phosphate for glycogen synthesis. Activation of glycogen synthesis will lead to a fall in the concentration of glucose 1-phosphate followed by a fall in glucose 6-phosphate, leading to an increase in hexokinase activity. Thus the flux from glucose to glycogen through hexokinase can be regulated independently of the glycolytic flux which is regulated by phosphofructokinase. This is essential since glycogen synthesis tends to occur when glycolytic flux is low.

14C4 Glucose transport

As described in Section 1D4, glucose transport in muscle is far from equilibrium and can be shown to be regulated. The regulation of hexokinase and phosphofructokinase is essentially by intrinsic effectors, but glucose transport responds to both intrinsic and extrinsic factors. Transport processes are more difficult to measure than enzyme-catalysed chemical reactions, and in this case, since glucose is rapidly converted to glucose 6-phosphate by hexokinase, it can be difficult to distinguish between effects on transport and effects on glucose metabolism. Two approaches can be used: the measurement of changes in intracellular glucose concentration and the measurement of uptake of analogues of glucose, such as arabinose or 3-O-methyl glucose, which are transported on the same carrier but not phosphorylated by hexokinase. This provides a direct measure of transport independent of metabolism.

Glucose transport is activated by insulin, anaerobic conditions, respiratory poisons such as cyanide, and uncouplers of respiratory chain phosphorylation. If fatty acids are available, the rate of glucose transport is also decreased. Thus the entry of glucose into the cell is increased by treatments which reduce the concentration of ATP, and decreased if an alternative fuel (fatty acid) is supplied. Insulin levels in the circulation rise in response to an increase in blood glucose. Insulin will therefore act as a signal of the availability of glucose for either glycolysis or glycogen synthesis. Insulin causes a marked increase in the intracellular glucose concentration indicating that glucose phosphorylation by hexokinase has become the main rate-controlling step for entry of glucose into either pathway. As described above, the regulation of hexokinase by glucose 6-phosphate allows the enzyme to respond to an increase in either glycolysis or glycogen synthesis. The effects of insulin are probably most important when the muscle is at rest. Under these conditions the stimulation of glucose entry allows glycogen synthesis, which is also directly stimulated by insulin, to take place (see Section 14F).

The mechanisms of regulation of glucose transport in muscle are obscure. In adipose tissue insulin is thought to increase the number of glucose transporters in the plasma membrane (see Chapters 10 and 16). It seems unlikely that different tissues will have different mechanisms, but there is no direct evidence for this mechanism in muscle. Extracellular free fatty acids are thought to inhibit the glucose transporter directly in adipose tissue, so again a similar mechanism might be expected in muscle. In the heart, such a mechanism would be physiologically valuable. The mechanism by which a fall in ATP activates glucose transport is unknown.

14C5 Control of glycolysis: conclusions

Muscle glycolysis is largely regulated in response to the energy demands of the cell, in other words by intrinsic controls. Extrinsic control operates to determine the balance between glucose and fatty acids in supplying the ATP demands of the cell, and the use of glucose for glycogen synthesis. There are direct hormone effects, such as the acceleration of glucose transport by insulin, and indirect effects, such as the response to the presence of fatty acids. The availability of fatty acids is largely controlled by hormones, adrenaline stimulating and insulin inhibiting fatty acid release from adipose tissue. In extreme situations the intrinsic controls will override the extrinsic controls. For example, if the muscle becomes anaerobic, glycolysis will be activated and this will not be affected by the presence of hormones or alternative substrates.

14D Regulation of glycogen metabolism

The reactions of glycogen synthesis and breakdown are shown in Fig. 14.6. In muscle, the only role of glycogen is to provide a rapidly available reserve of energy. There is no glucose 6-phosphatase in muscle to generate glucose from glucose 6-phosphate, so muscle glycogen cannot act as a source of circulating glucose. The energy for moderate exercise can be supplied from the oxidation of glucose or lipid from the circulation. As the work-load is increased, glycogen may become important as the energy demand exceeds the rate of entry of exogenous substrates into the cell. If the energy demand is sufficient to exceed the capacity of the muscle for oxidative metabolism, glycogen is mobilized very rapidly indeed to fuel anaerobic glycolysis. Under these conditions the glycogen resources are sufficient to support contraction for one or two minutes. White muscle has a poor capacity for oxidative metabolism. There are few mitochondria and the supply of oxygen through the circulation is relatively poor. Both the amounts of glycogen present and the capacity to mobilize glycogen tend to be greater than in red muscle.

Glycogen metabolism responds to two types of control. Breakdown is increased in response to an increased energy demand resulting from muscle contraction, an essentially intrinsic control. Adrenaline activates

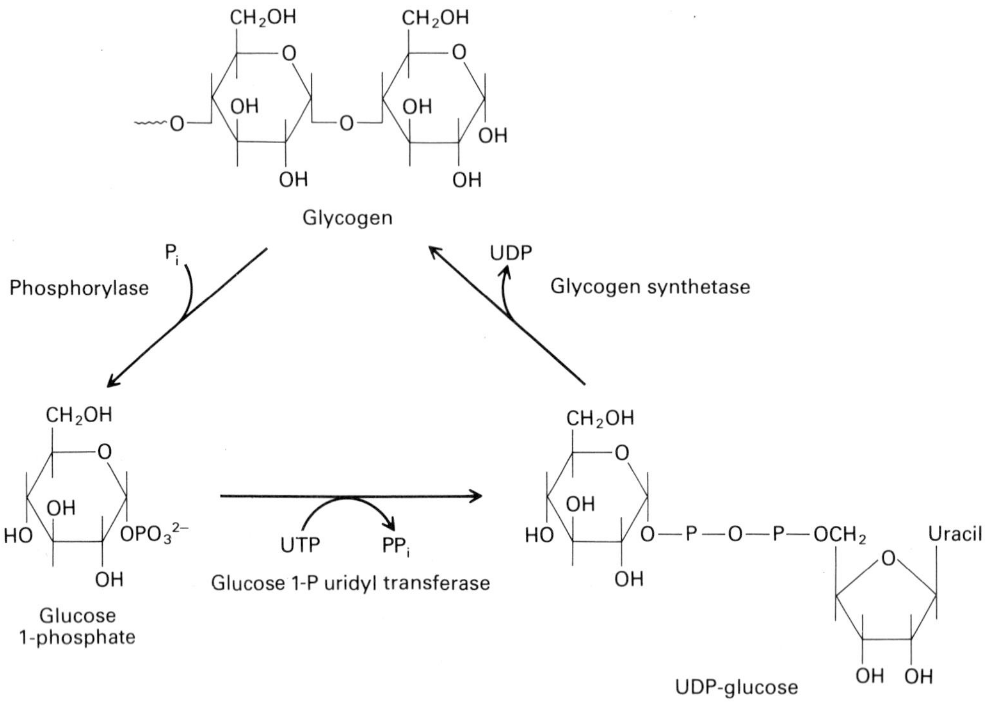

Fig. 14.6. Reactions of glycogen synthesis and breakdown. The steps regulated are glycogen synthetase and phosphorylase.

breakdown in resting muscle which mobilizes glycogen in preparation for contraction, a valuable response if, for example, the animal is faced by a predator. The synthesis of glycogen is largely controlled by insulin. Thus both synthesis and breakdown are subject to extrinsic control. Glycogen metabolism is regulated by a complex system of enzyme phosphorylations and dephosphorylations acting upon phosphorylase and glycogen synthetase (UDP-glucose glucosyl transferase) (Fig. 14.6).

14E Glycogen breakdown

14E1 The glycogenolytic cascade

Adrenaline stimulates glycogen breakdown by the cascade mechanism shown in Fig. 14.7. Amplification occurs at three separate steps: the activation of adenylate cyclase by adrenaline, the phosphorylation and

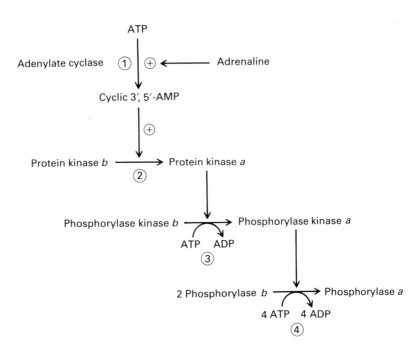

Fig. 14.7. The glycogenolytic cascade (in response to binding of adrenaline). Amplification takes place at steps 1, 3 and 4. Cyclic AMP activates protein kinase on a one-to-one basis, so no amplification occurs here.

activation of phosphorylase kinase, and the phosphorylation and activation of phosphorylase. This results in a large amplification of the original signal, the binding of adrenaline to its receptor. The activation of adenylate cyclase and cyclic AMP-dependent protein kinase was discussed in Chapters 3 and 4.

14E2 Phosphorylase kinase

Phosphorylase kinase is a complex enzyme containing four different types of subunit, α, β, γ and δ, with the overall composition $(\alpha, \beta, \gamma, \delta)_4$. The catalytic site is on the γ subunit. The α, β and δ subunits mediate the activation of the enzyme by phosphorylation and by Ca^{2+}.

The δ subunit is calmodulin and provides Ca^{2+}-binding sites. For each complex there will be a total of sixteen sites, four for each calmodulin (see Chapter 7). Calmodulin remains associated with the complex in the absence of Ca^{2+} ions which is unusual since in cytoplasmic enzymes the association usually depends upon the presence of Ca^{2+}. The activity of the enzyme depends upon Ca^{2+} binding to the δ subunit under all circumstances but the binding does not itself activate the enzyme very much. Two additional factors control the activity: phosphorylation by the cyclic AMP-dependent protein kinase to convert the b form of the enzyme to the a form, and further Ca^{2+} activation.

Phosphorylase kinase is phosphorylated by the cyclic AMP-dependent protein kinase on both the α and β subunits. The β-phosphorylation is more rapid and accounts for most of the activation effect. The slower α-phosphorylation may cause some additional activation. Phosphorylation causes a large increase in V_{max} and also increases the sensitivity to Ca^{2+} activation through the δ subunit (Fig. 14.8). Activity is still, however, almost completely dependent upon some Ca^{2+} being present.

The dephosphorylated state of the enzyme phosphorylase kinase b can be activated about thirtyfold by Ca^{2+} acting through a second class of Ca^{2+}-binding proteins known as δ'. In red muscle, the δ' component may be additional soluble calmodulin binding to the complex in the more usual Ca^{2+}-dependent manner. Alternatively, it may be the closely related Ca^{2+}-binding protein, troponin C. The main role of troponin C is to mediate the Ca^{2+} activation of myosin light chain kinase and hence the Ca^{2+} activation of muscle contraction. Control through troponin C allows a coordinated control of both muscle contraction and glycogen breakdown, and it seems likely that in red muscle the Ca^{2+} regulation of phosphorylase kinase is mediated by troponin C. In white muscle the Ca^{2+} activation of myosin light chain kinase is mediated by calmodulin and calmodulin will be responsible for both Ca^{2+} effects on phosphorylase kinase. The phosphorylated *a* form of the enzyme is not

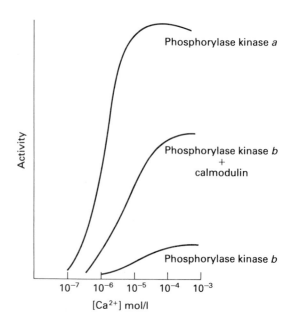

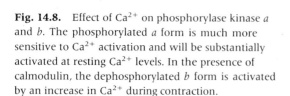

Fig. 14.8. Effect of Ca^{2+} on phosphorylase kinase *a* and *b*. The phosphorylated *a* form is much more sensitive to Ca^{2+} activation and will be substantially activated at resting Ca^{2+} levels. In the presence of calmodulin, the dephosphorylated *b* form is activated by an increase in Ca^{2+} during contraction.

further activated by the δ'/Ca^{2+} interaction. Thus two mechanisms of activation by hormones, cyclic AMP acting through the cyclic AMP-dependent protein kinase, or Ca^{2+} acting through binding proteins, offer alternative means of activating glycogen breakdown. In skeletal and heart muscle the major hormonal effect of adrenaline is mediated via β-receptors leading to an increase in cyclic AMP levels. In smooth muscle α-adrenergic effects are mediated by increased Ca^{2+}. Ca^{2+} is also important as an activator of glycogen breakdown during contraction.

14E3 Phosphorylation of phosphorylase kinase *in vivo*

In order to establish that phosphorylase kinase is activated by phosphorylation, it is necessary to show that the phosphorylation state changes in response to adrenaline *in vivo*. In principle the experiment is simple. An animal is injected with adrenaline and then sacrificed. The skeletal muscle is extracted in the presence of EDTA to remove Mg^{2+}, which inhibits protein kinases, and F^- ions, which inhibit protein phosphatases. Phosphorylase kinase is purified from the extract and the amount of covalently bound phosphate in the enzyme is determined. This can be compared with the enzyme from a control animal. Phosphorylase kinase *b* was found to contain substantial amounts of phosphate, about 2.3 moles per mole of (α, β, γ, δ) monomer. On treatment with adrenaline this only increased to 2.7 moles/mole of monomer. The phosphate in phosphorylase kinase *b* is probably irrelevant to the enzyme activity but it does provide a large background phosphate content so that the increase seen with adrenaline is not very convincing evidence of cyclic AMP-dependent phosphorylation *in vivo*.

The difficulty can be overcome by identifying the specific serine residues which are phosphorylated. This is done by using pure cyclic AMP-dependent protein kinase to phosphorylate pure phosphorylase kinase with $[^{32}P]ATP$. The phosphorylase kinase is then digested with trypsin and the peptides separated. Only two peptides were labelled, one in the α and one in the β subunit, thus there are only two sites subject to phosphorylation by the cyclic AMP-dependent protein kinase. In the whole animal, adrenaline can be shown to increase the phosphate content of these two sites. Furthermore, the activity of the enzyme correlates with the amount of phosphate in the site on the β subunit. In many phosphorylated enzymes it appears that much of the covalently bound phosphate is not involved in regulating the enzyme activity. The identification of specific sites, where the phosphorylation is changed in response to signal, is therefore necessary. The phosphorylation state of such sites can then be correlated with the activity of the enzyme.

14E4 Proteolysis of phosphorylase kinase

Phosphorylase kinase is activated by a short exposure to trypsin, yielding a form of the enzyme known as phosphorylase kinase a'. The activation is independent of phosphorylation. Muscle contains a Ca^{2+}-activated protease which has the same effect, so it is conceivable that Ca^{2+}-activated proteolysis might be a physiological mechanism for the activation of phosphorylase kinase. However, this seems unlikely since little phosphorylase kinase is found in the a' form in muscle extracts.

14E5 Phosphorylase

The final step of the glycogenolytic cascade is the conversion of phosphorylase b to phosphorylase a. Under most conditions this activates the enzyme. Phosphorylase b has two subunits with a molecular weight of 100 000 each. Phosphorylase kinase inserts a single phosphate into each subunit to give phosphorylase a which is found as a tetramer with a molecular weight of 400 000. Thus phosphorylation promotes dimerization of phosphorylase b. The increase in activity is much more rapid than the dimerization which is not therefore necessary for activation.

Phosphorylase b is regulated by the concentration of its substrate P_i and by activation by AMP. Increasing the concentration of AMP reduces the K_m for P_i, while increasing the P_i concentration reduces the K_a for AMP so that they have synergistic effect on the activity. The AMP effect is opposed by ATP and glucose 6-phosphate. Phosphorylase b is subject to intrinsic control by energy demand as reflected by changes in the concentration of P_i, AMP and ATP. In resting muscle, the relative concentrations are such that the enzyme has virtually no activity. On contraction, the ATP level falls slightly and the AMP and P_i rise substantially (see Section 14C1) activating the enzyme. The glucose 6-phosphate concentration acts as signal of glycolytic flux. Reduced energy demand leads to reduced glycolytic flux and increased glucose 6-phosphate concentration, brought about by reduction in the rate of phosphofructokinase (see Section 14C3).

Phosphorylase a is independent of AMP at high concentrations of P_i, but requires AMP at low concentrations of P_i. However, the K_a for AMP is much lower. Thus the effect of the covalent modification is to alter the response to direct binding effectors. This is common in phosphorylation-mediated regulation of enzymes. Under most conditions, conversion to phosphorylase a leads to substantial activation, but if phosphorylase b is already maximally activated by high levels of AMP and P_i, there will be relatively little effect. Phosphorylase is activated by three factors.

1 Energy demand expressed as rise in AMP and P_i, and a fall in ATP and glucose 6-phosphate. This acts upon phosphorylase *b*.
2 Muscle contraction mediated through the Ca^{2+}/troponin $C(\delta')$ activation of phosphorylase kinase causing the conversion of phosphorylase *b* to phosphorylase *a*. Muscle contraction is brought about by nervous stimulation.
3 Adrenaline working through the cyclic AMP-dependent phosphorylation and activation of phosphorylase kinase.

14F Glycogen synthesis

14F1 Glycogen synthetase

Glycogen synthetase catalyses the incorporation of glucose from UDP-glucose into glycogen. The enzyme has four identical subunits each with a molecular weight of 86 000. Early studies suggested that glycogen synthetase occurred in two forms: a phosphorylated-state glycogen synthetase *b* which requires glucose 6-phosphate for activity, and a dephosphorylated-state glycogen synthetase *a* which does not. Phosphorylation was thought to be catalysed by the cyclic AMP-dependent protein kinase so that a rise in cyclic AMP concentration tends to cause a reduction in activity. Later studies showed that the situation was more complicated and that glycogen synthetase is phosphorylated by at least five different protein kinases.
1 Cyclic AMP-dependent protein kinase phosphorylates sites 1a, 1b and 2.
2 Phosphorylase kinase phosphorylates site 2.
3 Glycogen synthetase kinase 3 phosphorylates sites 3a, 3b and 3c.
4 Glycogen synthetase kinase 4 phosphorylates site 2.
5 Glycogen synthetase kinase 5 phosphorylates site 5.

The activity of the enzyme is reduced by phosphorylation on sites 1a and 2, and 3a, b and c. This is known as multisite phosphorylation and in this case, as more sites are phosphorylated, the inhibition of the enzyme is progressively increased. This results from increases in the K_m for UDP-glucose and K_a for glucose 6-phosphate, and decreases in the K_I for ATP and P_i which antagonize glucose 6-phosphate activation. Glycogen synthetase does not therefore exist in two distinct states *a* and *b*, dephosphorylated and phosphorylated. Instead, the activity changes gradually as the phosphorylation state of a number of different sites changes.

Site 1b and site 5 do not appear to affect the activity of the enzyme directly. However, it appears that site 5 must be phosphorylated to allow the incorporation of phosphate into sites 3a, b and c by glycogen synthetase kinase 3.

14F2 Physiological role of regulation of glycogen synthesis

An effective control for glycogen breakdown in muscle requires only rapid and extensive activation in response to increased energy demand. The glycogenolytic cascade achieves this and also allows breakdown to respond to contraction through Ca^{2+} and the expected need for contraction through adrenaline. Glycogen synthesis, however, needs to respond to a number of different types of control.

1 To avoid wasteful cycling, synthesis needs to be inhibited under conditions where breakdown is activated.

2 The physical capacity of muscle cells to store glycogen is limited so the rate of synthesis needs to respond to the amount of glycogen present and to slow down when the maximum capacity is approached.

3 The rate of glycogen synthesis needs to respond to the availability of glucose in the circulation.

While the glycogenolytic response during contraction must be very rapid, the resynthesis of glycogen when the muscle returns to rest is less urgent. The activation of glycogen synthesis is relatively slow (Fig. 14.9) and the ability to respond to a number of different physiological factors is more important than the rate of response.

The prevention of wasteful cycling during contraction appears to operate largely through the phosphorylation of site 2. Both cyclic AMP-dependent protein kinase and phosphorylase kinase incorporate phosphate into this site inhibiting glycogen synthetase as they activate glycogen breakdown. Cyclic AMP-dependent protein kinase also incorporates phosphate into site 1a which further reduces activity. The role of glycogen synthetase kinase 4 which phosphorylates site 2 is not known.

The control of synthesis by the concentration of glycogen present in the muscle is less clear. In intact muscle it can be shown that glycogen synthetase becomes more phosphorylated as the concentration of glycogen increases. In fact, in resting muscle, glycogen synthetase tends to be in the phosphorylated inhibited state. However, in experiments using pure enzymes, glycogen does not affect the phosphorylation of glycogen synthetase by any of the protein kinases or its dephosphorylation by any protein phosphatases. The details of the mechanism are therefore obscure.

The availability of glucose is signalled by insulin which activates glucose transport. When energy demand is low and phosphofructo-

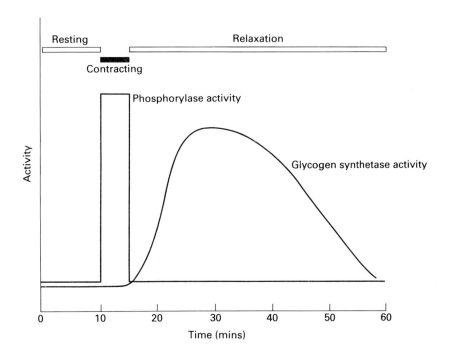

Fig. 14.9. Change in activity of phosphorylase and glycogen synthetase during contraction and relaxation. At rest, both enzymes are inhibited. On contraction, phosphorylase is very rapidly deactivated at the end of the contraction. Glycogen synthetase activity rises relatively slowly after the contraction and declines as the stores of glycogen are replenished.

kinase is inhibited, this leads to an increase in the concentration of glucose 6-phosphate which will tend to activate glycogen synthetase. However, there are effects of insulin on glycogen synthesis which are independent of glucose transport. Glucose transport *in vitro* can be increased by simply increasing the extracellular glucose concentration. This does not activate glycogen synthesis to anything like the same extent as adding insulin. Insulin does not affect the concentration of cyclic AMP in muscle or the extent of phosphorylation of sites 1 and 2. The hormone does cause a specific reduction in the phosphorylation of site 3 suggesting that it either inhibits glycogen synthetase kinase 3 or activates a site 3 specific phosphatase. None of the known protein phosphatases are specific for site 3 so inhibition of glycogen synthetase kinase 3 is the more likely explanation. Another possibility is regulation of glycogen synthetase kinase 5 whose only function appears to be to control the sensitivity of glycogen synthetase to phosphorylation by glycogen synthetase kinase 3. It is not known which, if any, of the proposed mechanisms of action of insulin are responsible for this effect.

The existence of several sites of phosphorylation on glycogen synthetase, many of which affect its activity, allows the enzyme to be regulated in response to several different physiological changes, each working through different sites and different protein kinases.

14G Protein phosphatases

A single enzyme, protein phosphatase 1, is responsible for the dephosphorylation of most of the phosphoprotein species in glycogen metabolism and probably many other physiological important phosphoproteins as well. Other protein phosphatases, 2A, B and C, have been identified. They all appear to have a fairly wide specificity although less so than phosphatase 1. Phosphatase 2B is activated by calmodulin but the physiological significance of this is unknown. It seems that protein dephosphorylation is less precisely regulated than protein phosphorylation, and that the protein phosphatases provide a background activity against which the protein kinases operate.

In muscle there is a mechanism by which a rise in cyclic AMP leads to inhibition of protein phosphatase 1. Muscle contains a phosphatase inhibitor protein which must itself be phosphorylated by the cyclic AMP-dependent protein kinase to be effective. Thus the reversal of cyclic AMP-dependent phosphorylation is inhibited. This makes it possible for adrenaline to activate phosphorylase substantially even in resting muscle. This raises the question of how the inhibitor protein is itself dephosphorylated when the protein phosphatase is blocked. The inhibitor is phosphorylated on a threonine residue rather than upon the usual serine residue. As a result it may still be susceptible to dephosphorylation by phosphatase 1. It may also be dephosphorylated by phosphatase 2.

Protein phosphatase activity may be affected by the state of the enzyme subject to dephosphorylation. For example, binding of AMP to phosphorylase *a* inhibits the removal of phosphate. It is possible that the binding of glycogen synthetase to glycogen inhibits its dephosphorylation providing a mechanism for the regulation of glycogen synthesis by the glycogen content of the cell. This is an attractive suggestion in view of the lack of any evidence for control of the rate of phosphorylation of glycogen synthetase by the glycogen. It is common for effectors which act by reversible binding to modulate the activity of an enzyme in two ways: by a direct effect and by altering the susceptibility of the enzyme to phosphorylation or dephosphorylation.

14H Glycogen metabolism: conclusions

An overall scheme for the regulation of glycogen metabolism is shown in Fig. 14.10. Both synthesis and breakdown are controlled by extrinsic regulators such as adrenaline and insulin, by muscle contraction through Ca^{2+} and by the demand for ATP and the supply of glucose. The many control mechanisms allow for the interplay of the different effectors in a coordinated manner. They can also accommodate the different requirements for the control of synthesis and breakdown.

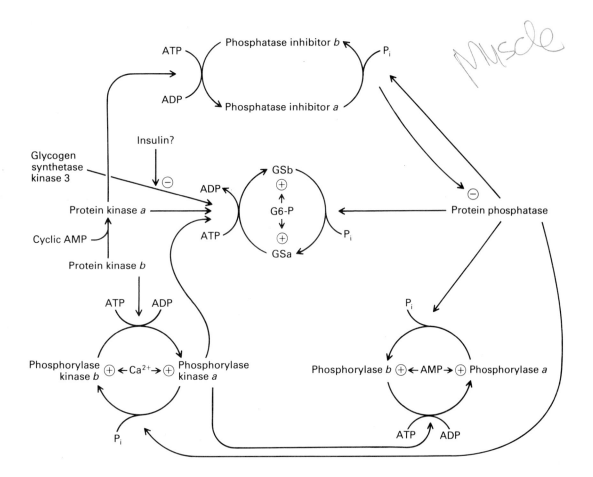

Fig. 14.10. Regulation of glycogen metabolism. This shows how cyclic AMP, acting initially through protein kinase, and Ca^{2+}, acting initially through phosphorylase kinase, affect phosphorylase and glycogen synthetase activity. For the sake of simplicity glycogen synthetase kinases 4 and 5, where function is unclear, are not shown.

14J Control of the citric acid cycle

Acetyl CoA provides the carbon source for oxidation and the generation of energy in the citric acid cycle. It may be derived from glucose via glycolysis and pyruvate dehydrogenase, from fatty acids by β-oxidation, or from ketone bodies. There is a need for the total acetyl CoA input to be regulated in relation to the energy demand and also for regulation of the source of acetyl CoA. In the presence of abundant fatty acids the oxidative metabolism of glucose is inhibited and so it might be expected that the production of acetyl CoA from pyruvate will also be inhibited. The rate of flow of the cycle itself needs to be regulated in relation to energy demand.

14J1 Regulation of pyruvate dehydrogenase

Pyruvate dehydrogenase is a large multienzyme complex located in the mitochondrial matrix. Overall it catalyses the reaction:

Pyruvate + NAD^+ + CoASH $\longrightarrow$ Acetyl CoA + NADH + H^+ + CO_2.

The complex contains three different enzymes which catalyse the reaction in the sequence shown in Fig. 14.11. Pyruvate dehydrogenase is subject to two levels of control. Both products, acetyl CoA and NADH, are inhibitors. The activity of the complex is also controlled by phosphorylation and dephosphorylation of the first enzyme in the sequence, pyruvate decarboxylase. The phosphorylated form of the enzyme is completely inactive. Pyruvate dehydrogenase in cell extracts can be phosphorylated and inhibited by cyclic AMP-dependent protein kinase, but in the intact cell, pyruvate dehydrogenase is not exposed to the cytoplasm and it is phosphorylated and dephosphorylated by a mitochondrial kinase and a mitochondrial phosphatase. These enzymes are apparently specific to pyruvate dehydrogenase, both being associated with the complex although the phosphatase is somewhat easier to remove than the kinase.

PDH kinase is inhibited by pyruvate and ADP and activated by acetyl CoA and NADH. It is not clear whether the effectors act directly upon the kinase itself, or upon the pyruvate dehydrogenase complex, altering the susceptibility of pyruvate decarboxylase to phosphorylation. This is difficult to determine since the kinase has no other known protein substrates. Abundant energy is reflected by high ratios of ATP to ADP, NADH to NAD and acetyl CoA to CoA. All will tend to increase the phosphorylation of pyruvate decarboxylase and the inhibition of pyruvate dehydrogenase activity. Fatty acid oxidation supplies acetyl

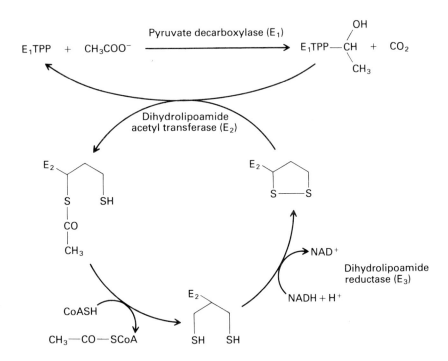

Fig. 14.11. Reactions of the pyruvate dehydrogenase complex.

CoA and NADH so a high level of β-oxidation will also tend to inhibit pyruvate dehydrogenase and the oxidation of glucose-derived carbon. This may be overcome to some extent if a lot of glucose is available since pyruvate inhibits pyruvate dehydrogenase kinase tending to activate the complex.

Pyruvate dehydrogenase phosphatase is activated by low levels of Ca^{2+}. This does not appear to involve a Ca^{2+}-binding protein but to be a direct effect. During muscle contraction, cytoplasmic Ca^{2+} rises and this can lead to a rise in the mitochondrial free Ca^{2+} (see Chapter 5) which may activate pyruvate dehydrogenase phosphatase and hence the pyruvate dehydrogenase complex.

In muscle, the phosphorylation state of pyruvate dehydrogenase is mainly controlled by intrinsic factors although the concentrations of the effectors may change in response to extrinsic effectors. In some tissues, however, hormones affect the phosphorylation state of PDH, the most notable example being the activation of PDH in adipose tissue by insulin (Chapters 10 and 16).

14J2 Regulation of lipid oxidation in muscle

Lipid is a major source of energy in red muscle and tends to be used in preference to glucose in aerobic conditions. However, the control of lipid metabolism in muscle is rather poorly understood. Red muscle does contain a certain amount of stored triacylglycerol. However, in a lean healthy individual the amounts are small and the activity of triacylglycerol lipase, which is required to mobilize these stores is low. In rat heart, for example, the activity is less than one per cent of the activity of phosphofructokinase. It seems likely, therefore, that the main source of lipid for oxidation in muscle is the circulation. Lipid-derived substrates are carried in the circulation in three forms: as free fatty acids bound to albumen, as triacylglycerol, in the form of lipoprotein, and as ketone bodies. The pathways of lipid oxidation are shown in Fig. 14.12.

A major factor in the control of lipid oxidation appears to be substrate supply. Fatty acids are mobilized from adipose tissue under conditions of exercise and starvation. During starvation, ketone body synthesis in the liver is accelerated. Exercise is associated with increased serum free fatty acid and starvation with an increase in both free fatty acids and ketone bodies. This is promoted by the action of glucagon and adrenaline on liver and adipose tissue, and inhibited by insulin, and will be discussed in detail in Chapters 15 and 16. It seems highly unlikely that regulation of the supply of circulating lipid is a sufficient control and there must be mechanisms to coordinate the rate of oxidation with the energy demands of the cell. The only convincing mechanism which has been suggested is limitation by the coenzyme A content of the mitochondria. If ATP levels are high, citric acid cycle flow is inhibited (see Section 14J3 below). This leads to a build up of acetyl CoA in the mitochondria. Since the total coenzyme A in the mitochondria is fixed the free coenzyme A falls so that less is available for β-oxidation. High levels of ATP in the mitochondria are associated with a high NADH/NAD$^+$ ratio which might also be expected to inhibit β-oxidation at hydroxyacyl-CoA dehydrogenase. High acetyl CoA levels might also be expected to inhibit ketone body oxidation. These suggestions are quite plausible and it is clear that the acetyl CoA level is important both as a regulator of lipid metabolism and of pyruvate oxidation, although the role is somewhat clearer in the liver (see Chapter 15). It seems unlikely, however, that lipid oxidation is regulated entirely by changes in the level of substrates in the circulation and by feedback inhibition by the end-product on an enzyme rather near the end of the pathway. However, if there are other sites of regulation they have yet to be identified.

14J3 Control of the citrate cycle itself

The citric acid cycle is shown in Fig. 14.13. It has been very difficult to

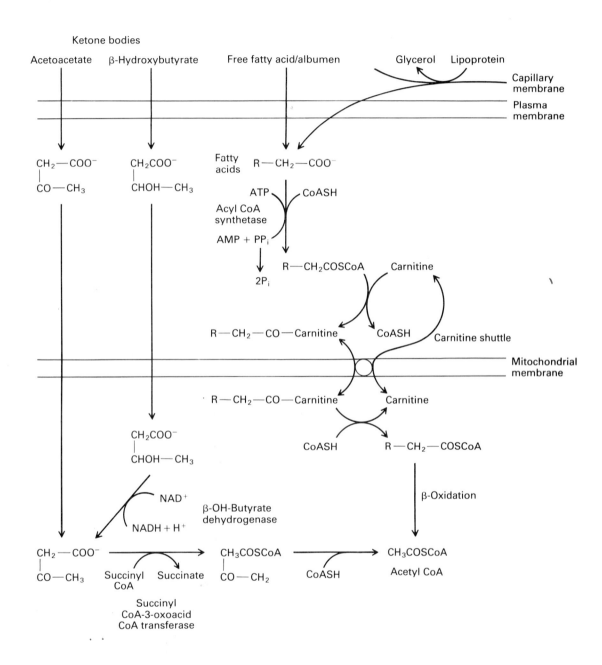

Fig. 14.12. Pathways for the conversion of circulating lipid to acetyl CoA.

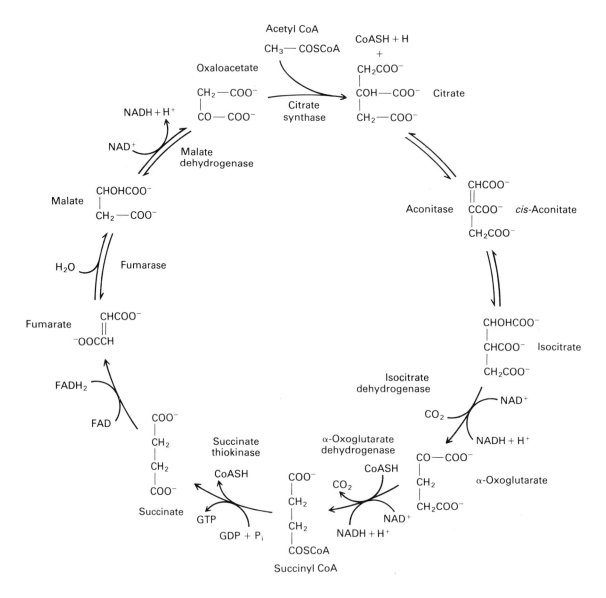

Fig. 14.13. Reactions of the citric acid cycle.

establish the pattern of regulation of the cycle and to identify the control sites. With the exception of succinyl CoA, all the intermediates of the citrate cycle occur in the cytoplasm as well as in the mitochondria. The distribution across the mitochondrial membrane is not known so it is impossible to determine the intramitochondrial concentration of citrate cycle intermediates in the intact cell. The mass action ratio method for identifying control sites is therefore not feasible.

An alternative approach is to study the properties of individual enzymes to see which might fit a regulatory role. In mammalian muscle three candidates emerge.

1 Citrate synthase is inhibited by ATP and citrate in competition with the substrates acetyl CoA and oxaloacetate.
2 NAD^+ isocitrate dehydrogenase is activated by Ca^{2+} and ADP, and inhibited by NADH and ATP.
3 α-Oxoglutarate dehydrogenase is inhibited by high $NADH/NAD^+$, ATP/ADP and succinyl CoA/CoA ratios, and activated by Ca^{2+}.

All these enzymes have relatively large negative free energy changes particularly α-oxoglutarate dehydrogenase. A number of observations are consistent with regulation at citrate synthase and α-oxoglutarate dehydrogenase. If lipid is supplied to a perfused heart, acetyl CoA levels rise rapidly and this is followed by a slower increase in citrate, isocitrate and α-oxoglutarate. The level of malate tends to fall. This suggests that the increase in acetyl CoA activates citrate synthase which must therefore be regulated. The build up of substrates round to α-oxoglutarate suggests that α-oxoglutarate dehydrogenase is rate limiting and therefore also a site of regulation. These conclusions are based on whole tissue concentrations of substrates, not mitochondrial concentrations. However, since lipid carbon can only be oxidized in the mitochondria it seems most unlikely that an increase in whole tissue concentration does not reflect an increase in the mitochondrial concentration.

14K Conclusions

The regulation of red muscle metabolism is subject to two physiological demands: the need to maintain the supply of ATP for contraction and the need to maintain blood glucose. The energy state of the cell is the main factor controlling metabolism, the signals being the relative concentrations of ATP, ADP and AMP, the $NADH/NAD^+$ ratio and the acetyl CoA/CoA ratio. Ca^{2+} acts as a signal of energy demand during concentration. Acetyl CoA acts as the main inhibitor of glucose oxidation when lipid is available acting directly at the level of PDH and indirectly through a rise in citrate at the level of phosphofructokinase.

Hormone regulation operates mainly to control the nature of the fuel supplied. Hormones directly affect the synthesis and mobilization of glycogen and the entry of glucose into the cell. Indirect effects of hormones result from changes in the supply of circulating lipid caused by hormone effects on other tissues.

Chapter 15   Regulation of metabolism in liver

15A Introduction

The major role of the liver is to maintain a fairly constant supply of circulating nutrients during variations in dietary intake. The maintenance of blood glucose levels during starvation is particularly important since the brain has an absolute requirement for glucose. The response to starvation is controlled by a rise in the circulating levels of glucagon and glucocorticoids and a fall in the level of insulin. Glucagon causes acute effects through the cyclic AMP-dependent protein kinase and, together with glucocorticoids, causes changes in the concentration of specific enzymes. These effects are antagonized by insulin. During short periods of fasting, liver glycogen is mobilized to release glucose. As fasting continues, gluconeogenesis from pyruvate, lactate, glycerol and amino acids begins. Gluconeogenesis from dietary protein will also occur if the diet is low in carbohydrate. Fatty acids cannot act as substrates for gluconeogenesis so a major energy store in the body cannot be used to maintain blood glucose and the animal must depend on proteins. During fasting the energy requirements of the liver are supplied by the oxidation of fatty acids from the circulation which will have been mobilized from adipose tissue. Fatty acids are also converted to ketone bodies which are released into the circulation and act as an alternative fuel. After lengthy periods of fasting the brain adapts to use ketone bodies which can supply about seventy per cent of its energy requirements.

In the well-fed state the direction of metabolism switches towards the storage of glycogen and the synthesis of triacylglycerol. Triacylglycerol is synthesized from fatty acids derived from circulating chylomicrons, and also from fatty acid synthesized in the liver. A little is stored in the liver but most is exported as lipoprotein for storage in adipose tissue. It used to be thought that the main carbon source for glycogen synthesis and fatty acid synthesis in the well-fed state was circulating glucose. Work done in the last ten years or so, has shown that glucose is a poor substrate for liver at any circulating concentration below about 12 mmol/l. In a healthy individual, as opposed to a diabetic, blood glucose rarely exceeds 12 mmol/l. It appears that gluconeogenic precursors are better substrates than glucose for both fatty acid biosynthesis and glycogen synthesis. In the fed state, the most important substrates quantitatively are likely to be lactate and, particularly in man, fructose. In the developed world, human diets are rich in sucrose and hence in fructose. It appears that the metabolism of glucose in liver is slow under most conditions. As the blood glucose increases, the metabolism of gluconeogenic substrates switches from the production of glucose towards glycogen synthesis, fatty acid synthesis, and also, oxidation through the citric acid cycle for energy production. The liver is also a major site for the biosynthesis of purines and steroids and the only organ able to synthesize urea. This chapter will concentrate on the regulation of the pathways which maintain a balance between carbohydrate and lipid metabolism.

15B Lipid metabolism

15B1 Fatty acid oxidation

The pathway of fatty acid odixation is shown in Fig. 15.1. After entry into the cell, fatty acids are esterified with coenzyme A. The pyrophosphate produced is rapidly hydrolysed to P_i, increasing the overall negative free energy of the reaction, so that the production of acyl CoA is favoured. The mitochondrial membrane is impermeable to fatty acyl CoA and the transfer of fatty acyl units into the mitochondria where β-oxidation takes place is achieved by the carnitine shuttle (Fig. 15.1). This involves the synthesis of acyl carnitine from acyl CoA and carnitine in the cytoplasm, and the regeneration of acyl CoA in the mitochondria by the reverse reaction. Acyl carnitine enters the mitochondria on a carrier in exchange for free carnitine. The cytoplasmic transferase is inhibited by malonyl CoA, which is an intermediate in fatty acid biosynthesis, formed in the cytoplasm by acetyl CoA carboxylase. In liver, acetyl CoA carboxylase is the main regulatory site for fatty acid synthesis (see Section 15B4), so that as synthesis is activated the malonyl CoA concentration might be expected to increase resulting in inhibition of fatty acid oxidation. The function of this control is probably to prevent wasteful cycling between synthesis and oxidation, rather than to regulate the oxidation of fatty acids taken up from the circulation.

The supply of free fatty acid in the circulation seems to be the main factor regulating β-oxidation in the liver, and an increase in serum free fatty acid concentration can be shown to increase the rate of β-oxidation. The level of circulating free fatty acid is determined mainly by the rate of lipolysis in adipose tissue. Lipolysis is activated by glucagon during starvation and by adrenaline during stress or exercise, both working through cyclic AMP. Insulin inhibits lipolysis and the fall in circulating insulin concentration is a major factor in increasing rates of lipolysis during starvation. Thus the rate of fatty acid oxidation is under hormonal control, but the effect is indirect being a consequence of hormone effects on a different tissue.

Increased fatty acid oxidation in liver leads to a marked increase in

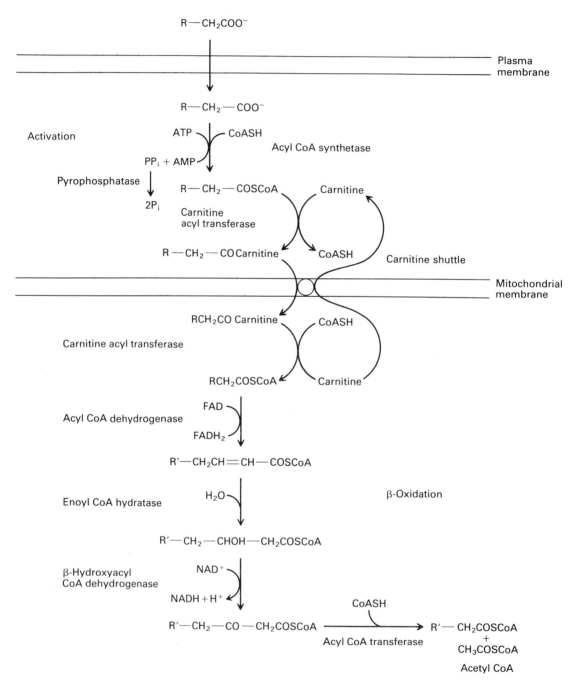

Fig. 15.1. Pathway of fatty acid oxidation. There are three stages. (1) Entry into the cell and activation by formation of the CoA ester. (2) Transfer into the mitochondria via the carnitine shuttle. (3) β-Oxidation, leading to the sequential removal of 2 carbons at a time as acetyl CoA.

the level of acetyl CoA in the mitochondria, and this is thought to be an important signal in the activation of gluconeogenesis and ketone body biosynthesis.

15B2 Ketone body synthesis

The ketone bodies, acetoacetate and β-hydroxybutyrate are synthesized in liver mitochondria by the pathway shown in Fig. 15.2. Of the three enzymes involved, acetoacetyl CoA thiolase, β-hydroxymethyl glutaryl CoA synthetase (HMG-CoA synthetase) and HMG-CoA lyase, HMG-CoA synthetase has the lowest activity and is thought to be rate limiting. The equilibrium constant of acetoacetyl CoA thiolase is equal to:

$[\text{acetyl CoA}]^2 / [\text{CoASH}][\text{acetoacetyl CoA}]$.

If the reaction is close to equilibrium, acetoacetyl CoA will rise in proportion to the square of the acetyl CoA concentration and at the same time the concentration of CoASH will be falling. Thus both substrates of HMG-CoA synthetase increase as the acetyl CoA concentration increases. During starvation, mitochondrial acetyl CoA levels rise as a result of increased fatty acid oxidation, and the resulting substrate stimulation of HMG-CoA synthetase is thought to be the main factor regulating ketone body synthesis. However, glucagon or dibutyryl cyclic AMP stimulate ketone body synthesis in isolated perfused liver. The concentration of glucagon required is higher than normally occurs *in vivo*. Nevertheless, glucagon does appear to be able to stimulate ketone body synthesis by a cyclic AMP-mediated mechanism which must be independent of changes in circulating free fatty acids, since the perfused liver is isolated from the circulation.

Recent studies of HMG-CoA synthetase have shown that the enzyme is inhibited by succinyl CoA. Inhibition results from succinylation of the enzyme at a site normally occupied by acetate during the reaction. Reversal of the inhibition is quite slow and is favoured by acetyl CoA. This provides an additional mechanism whereby an increase in acetyl CoA might activate ketone body synthesis. It may also provide an explanation for the direct glucagon effect since glucagon reduces the concentration of succinyl CoA in liver. Succinyl CoA occurs only in the mitochondria so HMG-CoA synthetase must be exposed to this change in concentration. The reasons for the fall in succinyl CoA in response to glucagon are not clear. It seems that ketone body synthesis may be activated by a rise in mitochondrial acetyl CoA levels, which is ultimately a consequence of hormonal stimulation of lipolysis in adipose tissue, and a fall in succinyl CoA resulting from a direct effect of glucagon on the liver cell.

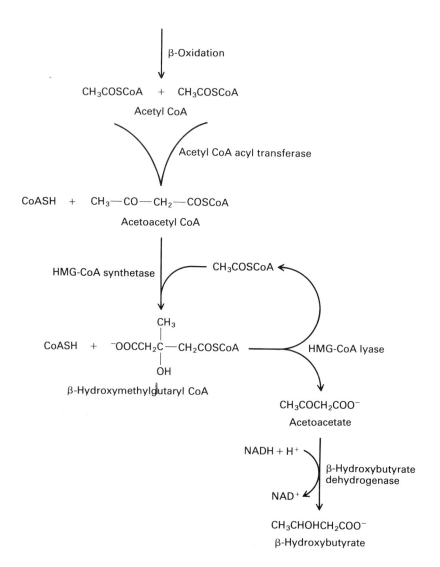

Fig. 15.2. Reactions of ketone body synthesis in liver. All reactions take place in the mitochondria. Acetoacetate and β-hydroxybutyrate are exported into the circulation.

15B3 Importance of ketone body synthesis

The stimulation of ketone body synthesis is largely a result of an increased rate of fatty acid oxidation, and the very high rates of fatty acid oxidation observed in liver during starvation depend upon ketone body synthesis. The mitochondrial membrane is impermeable to coenzyme A so that the total coenzyme A in the mitochondria is fixed. The synthesis of acetoacetate can be described by the equation:

2 acetyl CoA ⟶ acetoacetate + 2 CoASH (see Fig. 15.2).

Thus ketone body synthesis acts as a mechanism for regeneration of free CoASH. In the absence of ketone body synthesis, fatty acid oxidation would be limited by the capacity of the citric acid cycle to oxidize the acetyl CoA produced.

During prolonged starvation the level of acetoacetate and β-hydroxybutyrate in the circulation reaches about 8 mmol/l. This compares with a total free fatty acid concentration of about 2 mmol/l, but the concentration of free fatty acids in solution is only about 20 μmol/l, the rest being bound to albumen. Thus the synthesis and release of ketone bodies greatly increases the amount of circulating fuel derived from lipid. Heart muscle and kidney have particularly large capacities for ketone body utilization and during starvation will obtain the bulk of their energy from the oxidation of ketone bodies (see Section 14J2). During prolonged starvation the brain adapts to replace about seventy-five per cent of its glucose requirement with ketone bodies.

In diabetes the concentration of ketone bodies in the circulation may be as high as 25–30 mmol/l. The reason for this is not entirely clear. Diabetics do have high rates of lipolysis and high levels of fatty acid in the serum, but the fatty acid concentrations are no higher than in prolonged starvation, while the ketone body levels are three times as high. This provides further support for the view that there are direct effects of hormones on ketogenesis, and suggests that insulin may exercise an inhibitory effect in addition to the inhibition resulting from a reduction in the availability of free fatty acids. However, there is no direct evidence for such an effect. The high levels of acetoacetate and β-hydroxybutyrate in diabetes lead to an acidification of the blood. Many of the clinical problems associated with diabetes arise from this acidosis.

15B4 Fatty acid biosynthesis

Fatty acid biosynthesis has been extensively studied in rat fat cells and this will be described in detail in the next chapter. In fat there are several sites of control; in liver the main site for rapid regulation is acetyl CoA carboxylase, the first enzyme which is exclusively involved in fatty acid biosynthesis (Fig. 15.3). The enzyme is activated by insulin and inhibited by adrenaline. The mechanism is the same as in adipose tissue and is discussed in the next chapter (Section 16C). Control at this point makes physiological sense, since the main carbon sources for fatty acid biosynthesis in liver are fructose and lactate, in contrast to fat cells where the main carbon source is glucose.

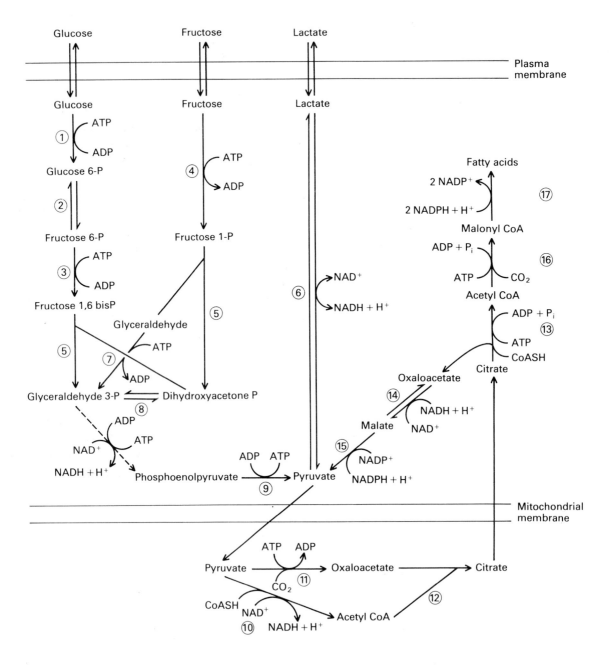

Fig. 15.3. Fatty acid biosynthesis in liver. The substrates are glucose, fructose and lactate. In a healthy, i.e. non-diabetic, individual, fructose and lactate are the most important. The reactions involved are: (1) glucokinase, (2) glucose 6-P isomerase, (3) phosphofructokinase, (4) fructokinase, (5) aldolase, (6) lactate dehydrogenase, (7) triose kinase, (8) triose phosphate isomerase, (9) pyruvate kinase, (10) pyruvate

dehydrogenase, (11) pyruvate carboxylase, (12) citrate synthase, (13) ATP-citrate lyase, (14) NAD-malate dehydrogenase, (15) NADP-malate dehydrogenase (malic enzyme), (16) acetyl CoA carboxylase, (17) fatty acid synthetase.

15C The citric acid cycle

In liver, the citric acid cycle has two roles. The whole cycle may be used to oxidize acetyl CoA for the generation of ATP. The span of the cycle from α-oxoglutarate to oxaloacetate is also used for the interconversion of carbon skeletons for synthesis. Gluconeogenesis requires either malate or oxaloacetate in the mitochondria, depending upon the location of phosphoenolpyruvate carboxykinase (see Section 15E3). The second half of the citric acid cycle is also used to convert the carbon skeletons of many amino acids into gluconeogenic intermediates (Fig. 15.4).

The two functions of the citrate cycle can operate as essentially independent pathways. Acetyl CoA oxidation depends upon the availability of oxaloacetate for the formation of citrate, and this depends upon the operation of the second section of the cycle. However, as long as the level of oxaloacetate is high enough to maintain citrate synthase activity, any additional flux in the span from α-oxoglutarate to oxaloacetate can be used for other purposes. In gluconeogenic conditions there may be net flux from oxaloacetate to malate, the reverse of the usual direction of flow in the citric acid cycle.

The rate of oxidation of acetyl CoA through the citrate cycle in liver is much lower than the rate in muscle under all conditions. If there is abundant carbohydrate in the diet there will be a high rate of fatty acid synthesis. One of the major carbon sources will be fructose. Fatty acid synthesis from fructose or glucose results in net synthesis of ATP (see Section 16B), so the requirement for ATP from the citrate cycle is relatively low. Lactate, or amino acids, may also provide carbon for fatty acid synthesis in the well-fed state. In this case the energy required is presumably supplied by oxidizing a proportion of the lactate or amino acids via the citrate cycle. Under these conditions the citrate cycle is presumably controlled by the energy requirement, in a similar way to muscle (see Section 14J3). However, the energy demands of liver cells are always much less than the energy demands of contracting muscle. Another point is, that while the change in energy requirement from resting to contracting muscle is very large, the energy demand in liver does not change very much, whatever the dietary state. The substantial differences are in the carbon source used to generate ATP rather than in the amount of ATP required.

During starvation, or with a low carbohydrate content in the diet,

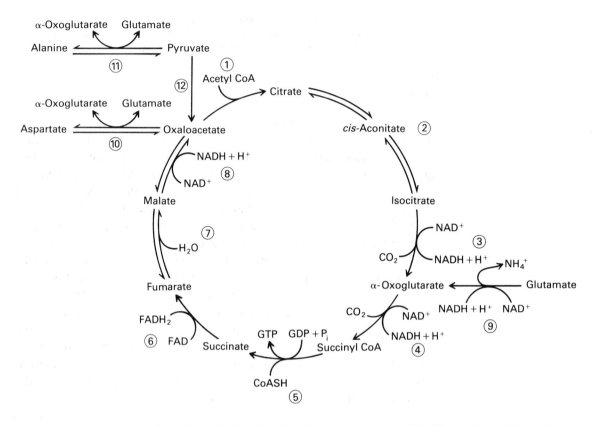

Fig. 15.4. Citric acid cycle in liver: connections with amino acid metabolism. The reactions involving alanine, aspartate and glutamate, feeding into the citrate cycle and then to oxaloacetate for gluconeogenesis, are shown. Most other amino acids also yield four carbon skeletons which can be converted to oxaloacetate via the citrate cycle. The reactions involved are: (1) citrate synthase, (2) aconitase, (3) isocitrate dehydrogenase, (4) α-oxoglutarate dehydrogenase, (5) succinate thiokinase, (6) succinate dehydrogenase, (7) fumarase, (8) malate dehydrogenase, (9) glutamate dehydrogenase, (10) aspartate aminotransferase, (11) alanine aminotransferase, (12) pyruvate carboxylase.

the rate of oxidation of acetyl CoA through the citric acid cycle is even lower, although the rate of flow in the span from α-oxoglutarate to malate, and oxaloacetate to malate, increases. During starvation, rates of fatty acid oxidation are high and there is a marked increase in the level of acetyl CoA. Acetyl CoA is a substrate for citrate synthase so the increase in concentration suggests that this is the step which limits oxidation of acetyl CoA through the citric acid cycle.

Three factors have been suggested to explain the reduction in flow through citrate synthase during the onset of starvation: a reduction in oxaloacetate concentration, an increase in ATP, or an in[creased] concentration of fatty acyl CoA. The high rate of β-oxidation [is] expected to increase the NADH/NAD$^+$ ratio and, as a result, th[e ATP/] ADP ratio. ATP has been proposed as a direct inhibitor of citrate synth[ase], and an increase in the NADH/NAD$^+$ ratio would be expected to mov[e] the equilibrium of malate dehydrogenase in favour of malate, and so reduce the concentration of oxaloacetate (see Fig. 15.5). At the same time, the level of fatty acyl CoA might be expected to increase as a consequence of increased rates of entry of fatty acids into the cell during starvation.

The hypothesis is attractive, and manipulating the ATP content and redox state of isolated liver mitochondria does result in inhibition of the citrate cycle. However, during prolonged starvation the total cell ATP falls. It is just possible that the mitochondrial ATP rises and that this is compensated for by a larger fall in cytoplasmic ATP, but this seems unlikely. A rise in mitochondrial ATP and NADH would also inhibit α-oxoglutarate dehydrogenase and hence gluconeogenesis from glutamate, which yields oxoglutarate by transamination or by the glutamate dehydrogenase reaction (see Fig. 15.5). Furthermore, citrate synthase is inhibited by ATP^{4-}, and not by MgATP^{2-}, which is the major form in which ATP occurs in the cell.

During gluconeogenesis, the concentrations of malate and aspartate fall. Both are thought to be in equilibrium with oxaloacetate through the malate dehydrogenase and aspartate aminotransferase reactions (Fig. 15.5), so it is likely that the concentration of oxaloacetate falls as well. The mechanism behind a fall in oxaloacetate is not known. It might result from activation PEP-CK which is thought to be a rate-limiting enzyme for gluconeogenesis and has oxaloacetate as a substrate. The activity of PEP-CK increases during starvation. Alternatively, a rise in the mitochondrial NADH/NAD ratio would lead to the conversion of

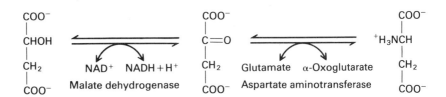

Fig. 15.5. Interconversions between malate, oxaloacetate and aspartate. Both reactions are likely to be close to equilibrium in the cell, so that a reduction in oxaloacetate concentration might be expected to be reflected in a reduction in the concentration of malate and aspartate.

te to malate. In rat liver, where phosphoenolpyruvate car-
 is exclusively cytoplasmic, gluconeogenesis does involve
 of oxaloacetate to malate in the mitochondria (see Section
 ibition of citrate synthase by fatty acyl CoA is also plausible
 acyl CoA levels rise in liver cells during starvation.

ycogen metabolism in liver

Liver glycogen provides a reserve of glucose which is sufficient to cope with day-to-day variations in food intake, but not to maintain blood glucose during an extended period of fasting. In muscle, maximal stimulation of glycogenolysis leads to the glycogen reserves being exhausted in one or two minutes. In liver, glycogen normally lasts for about 24 hours, and under the most extreme conditions of prolonged exercise will last for about 90 minutes.

The control of liver glycogen breakdown is summarized in Fig. 15.6. The signals are the same as in muscle but the nature of the response is different, since changes in the energy state of the cell are not very important. During the first 24 hours of fasting, when glycogen mobilization is important, ATP and AMP levels in the liver do not change to any significant extent and regulation by hormones is much more significant. Glucagon activates glycogenolysis through the cyclic AMP-mediated cascade. Vasopressin, angiotensin and adrenaline, acting through α_1-adrenergic receptors, raise the cytoplasmic free Ca^{2+} which activates phosphorylase kinase and increases glycogenolysis (see Chapter 6). Thus in liver, Ca^{2+} acts primarily as a hormone second messenger. During glycogenolysis, glycogen synthesis will be inhibited by the phosphorylation of glycogen synthetase by the cyclic AMP-dependent protein kinase, and by phosphorylase kinase preventing cycling between synthesis and breakdown as in muscle.

The regulation of glycogen synthesis is less well understood and the protein kinases involved are less well characterized than in muscle. Synthesis is activated by increased blood glucose, and this is probably mediated by an increase in the level of glucose 6-phosphate, although the main carbon sources are probably fructose and lactate. Insulin also activates glycogen synthesis and inhibits glycogen breakdown. As in muscle, the mechanisms are imperfectly understood. In liver, insulin may antagonize the effects of glucagon on glycogen metabolism by virtue of its ability to lower cyclic AMP levels in the presence of glucagon (see Chapter 10), but this cannot explain the activation of synthesis in the absence of glucagon. This probably results from a reduction in the phosphorylation of glycogen synthetase, but the details have not been established.

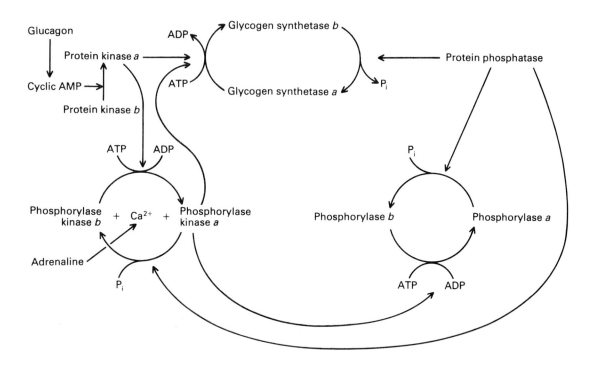

Fig. 15.6. Regulation of glycogen metabolism in liver. Glycogen breakdown is activated by glucagon acting through a rise in cyclic AMP, and by adrenaline acting through a rise in Ca^{2+}-activating phosphorylase kinase.

15E Glycolysis and gluconeogenesis in liver

Liver has the capacity to utilize glucose via glycolysis, or synthesize glucose via gluconeogenesis. Under most dietary conditions rates of glucose utilization are low. Substantial rates of glycolysis from glucose in liver require blood glucose levels in excess of 20 mmol/l, much higher than the concentrations encountered in a healthy individual. Thus the control of the two pathways, glycolysis and gluconeogenesis, is mainly concerned with controlling the net rate of gluconeogenesis. Another important requirement is to determine the fate of glucose 6-phosphate produced by gluconeogenesis, which may be used for glycogen synthesis or for glucose release.

The pathway of gluconeogenesis is largely a reversal of the pathway of glycolysis (Fig. 15.7). However, three steps in glycolysis have a large negative free energy change and are considered to be irreversible: the

256 *Chapter 15*

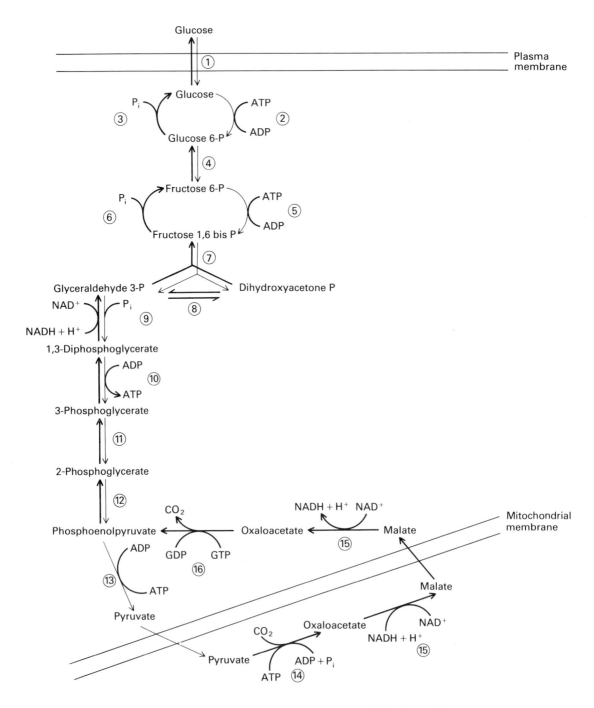

Fig. 15.7. Pathways of gluconeogenesis and glycolysis. Gluconeogenesis is shown by the bold continuous line and glycolyis by the fine black line. The reactions involved

are: (1) glucose transport, (2) glucokinase, (3) glucose 6-phosphatase, (4) glucose 6-phosphate isomerase, (5) phosphofructokinase, (6) fructose 1, 6-bisphosphatase, (7) aldolase, (8) triose phosphate isomerase, (9) glyceraldehyde 3-P isomerase, (10) phosphoglycerate kinase, (11) phosphoglycerate isomerase, (12) enolase, (13) pyruvate kinase, (14) pyruvate carboxylase, (15) malate dehydrogenase, (16) phosphoenolpyruvate carboxykinase.

conversion of glucose to glucose 6-phosphate, fructose 6-phosphate to fructose 1,6-bisphosphate and phosphoenolpyruvate to pyruvate. Gluconeogenesis requires enzymes which will catalyse carbon flow in the reverse direction at all these steps, forming substrate cycles. Obviously the substrate cycles are likely to be the sites where the pathway is regulated.

15E1 Glucose transport and phosphorylation

Liver is able to either take up or release glucose, depending upon the dietary state. For this to be possible the transport process must be close to equilibrium, in contrast to the situation in adipose tissue and muscle. Since the equilibrium constant for glucose transport is one, this means that the cytoplasmic glucose concentration is close to the circulating concentration, and that both rise and fall together. It follows that glucose transport is never rate limiting and cannot function as a site of control. This reflects the fact that the maximum rate of transport is much greater than the maximum rate of phosphorylation to glucose 6-phosphate which in liver is the first rate-controlling step for glycolysis.

In liver, the conversion of glucose to glucose 6-phosphate is catalysed by glucokinase rather than hexokinase. Hexokinase is present but the activity is very low. Hexokinase has a low K_m for glucose of about 50 µmol/l. In muscle, or adipose tissue, glucose transport is rate controlling, and the low K_m and high V_{max} of hexokinase results in a low intracellular glucose concentration, glucose being trapped as glucose 6-phosphate. Glucokinase in liver has a high K_m for glucose and a relatively low V_{max}. As a result, glucose transport is not rate limiting in liver, and intracellular glucose concentrations are close to the extracellular glucose concentration. The reverse reaction is catalysed by glucose 6-phosphatase.

Glucokinase is only found in the liver, and glucose 6-phosphatase in the liver and kidney cortex, the only other gluconeogenic tissue. The two enzymes form a substrate cycle. Both reactions have a large negative free energy change, and both appear to be rate controlling since their mass action ratio in the cell is far removed from equilibrium. Since this substrate cycle controls the direction of flow either towards or away from glucose, both enzymes might be expected to be subject to a high degree of regulation. In fact it appears that the main factor in both cases is the substrate concentration. Glucokinase has a K_m for glucose of about 10 mmol/l. Blood glucose in a well-fed individual before a meal is about 5 mmol/l, and after a meal rises to about 12 mmol/l. The intracellular glucose will rise to the same extent causing an increase in the rate of flux to glucose 6-phosphate through glucokinase. The increase will be quite small. Since the K_m for glucokinase is 10 mmol/l, an increase in glucose concentration from 5 mmol/l to 12 mmol/l will increase the rate of flux through glucokinase by about twofold. Nevertheless, this seems to be the only rapid control that operates. The reverse reaction glucose 6-phosphatase also appears to have no regulators other than substrate concentration.

The regulation of blood glucose maintains a concentration of 5 mmol/l in a well-fed individual. At this concentration there will be no net flux of glucose into the liver, and little net flux out of the liver. However, since this approximates to about half the K_m for glucokinase, the enzyme will be operating at approximately twenty-five per cent of its maximal velocity. There must therefore be an approximately equal flux in the opposite direction through glucose 6-phosphatase, implying a substantial rate of cycling between glucose and glucose 6-phosphate. If we assume that the flux through glucose 6-phosphatase remains constant, a fall in blood glucose will cause a reduction in the rate through glucokinase resulting in increased net glucose release. A rise in blood glucose will cause an increase in the flux through glucokinase, ultimately resulting in net glucose uptake. The activity of glucokinase is such that relatively high circulating glucose concentrations are needed to produce significant net uptake. This very simple mechanism has the advantage that it responds in a very direct way to changes in blood glucose concentration, acting in effect as a glucose buffer.

The limited ability of substrate control to produce a change in rate through the individual enzymes is compensated for by the amplification effect of a high rate of cycling between glucose and glucose 6-phosphate. The situation is analogous to the operation of the phosphofructokinase/fructose bisphosphatase substrate cycle in muscle (see Chapters 1 and 14). In that case the major pathway is glycolysis, and the gluconeogenic enzyme, fructose bisphosphatase, provides a mechanism for increasing the sensitivity of control of glycolytic flux. In the liver, the main pathway is gluconeogenesis and the glycolytic enzyme, glucokinase, provides a mechanism for increasing the sensitivity of control of glucose release. The flux through glucose 6-phosphatase will change in response to the cellular glucose 6-phosphate concentration. Glucose 6-phosphate is close to equilibrium with glucose 1-phosphate and fructose 6-phosphate, and will rise if the glucose 1-phosphate increases due to stimulation of glycogen breakdown, or if fructose 6-phosphate increases as a result of inhibition of phosphofructokinase or activation of fructose 1,6-bisphosphatase.

In the long term the concentrations of the enzymes may change. During starvation, the concentration of glucose 6-phosphatase increases and the concentration of glucokinase decreases, presumably in response to glucagon and glucocorticoids. This alters the balance of the substrate cycle in favour of glucose production. Insulin has the opposite effect.

15E2 The fructose 6-phosphate/fructose 1,6-bisphosphate substrate cycle

The next irreversible step in glycolysis is the conversion of fructose 6-phosphate to fructose 1,6-bisphosphate by phosphofructokinase. The corresponding gluconeogenic enzyme is fructose bisphosphatase. Unlike glucose 6-phosphatase, which only occurs in liver and kidney cortex, the two gluconeogenic tissues, fructose bisphosphatase is also found in muscle. The role of the substrate cycle in muscle is to allow amplification of the glycolytic response to ATP demand (see Secton 14C2). The role in liver is to allow for net flow in the gluconeogenic direction. Under most conditions net flux is in the gluconeogenic, rather that the glycolytic, direction. It is not surprising then, that in liver the V_{max} activity of fructose bisphosphatase is three or four times higher than the V_{max} of phosphofructokinase, while in muscle it is only about ten per cent of the phosphofructokinase activity.

The regulation of the two enzymes in liver and muscle appears to be very similar. Phosphofructokinase is inhibited by ATP, and the inhibition is relieved by AMP, P_i, and fructose 1,6-bisphosphate and increased by citrate. Fructose bisphosphatase is strongly inhibited by AMP. From this it appears that when the cell has abundant energy, glucose synthesis is activated. It is unlikely that this mechanism operates *in vivo*. During prolonged starvation, when gluconeogenesis is important, levels of ATP in liver tend to fall rather than rise. During shorter periods of fasting, ATP may rise or fall depending upon the precise circumstances. In any case, rates of gluconeogenesis correlate poorly with the concentration of AMP and ATP. The mechanism might become important during anoxia, activating glycolysis, the only source of ATP duirng anaerobic conditions, and inhibiting gluconeogenesis.

Recently a second regulatory mechanism has been discovered.

Fructose 2,6-bisphosphate has been found to activate phosphofructokinase. It relieves the inhibition by ATP, and also decreases the K_m for fructose 6-phosphate. It is a powerful inhibitor of fructose bisphosphatase, increasing the K_m for fructose 1, 6-bisphosphate. As well as affecting both enzymes directly, fructose 2,6-bisphosphate potentiates both the activatory effect of AMP on phosphofructokinase and the inhibitory effect of AMP on fructose bisphosphatase.

Fructose 2,6-bisphosphate is formed by the reaction:

fructose 6-phosphate + ATP ⟶ fructose 2,6-bisphosphate + ADP

which is catalysed by phosphofructokinase 2, a different enzyme from phosphofructokinase 1, the enzyme in the mainstream of glycolysis, which converts fructose 6-phosphate to fructose 1,6-bisphosphate. Fructose 2,6-bisphosphate is converted to fructose 6-phosphate by fructose bisphosphatase 2 (fructose 2,6-bisphosphatase), which is a different enzyme from fructose bisphosphatase 1, which converts fructose 1,6-bisphosphate to fructose 6-phosphate. Both activities appear to be on a single enzyme, which is phosphorylated by the cyclic AMP-dependent protein kinase. Phosphorylation inhibits phosphofructokinase 2 activity and stimulates fructose bisphosphatase 2 activity. Therefore a rise in cyclic AMP causes a reduction in the concentration of fructose 2,6-bisphosphate leading to inhibition of phosphofructokinase 1 and activation of fructose bisphosphatase 1. This results in the activation of gluconeogenesis. In liver, adenylate cyclase is most responsive to the main gluconeogenic hormone, glucagon. This mechanism constitutes a cascade through which glucagon can activate gluconeogenesis and inhibit glycolysis, in spite of the fact that ATP levels are falling and AMP levels rising (Fig. 15.8). Thus the main control on the fructose 6-phosphate/fructose 1,6-bisphosphate substrate cycle is a direct cyclic AMP-mediated hormone effect, which operates to override the intrinsic controls on the system resulting from changes in the concentration of ATP.

15E3 The interconversion between pyruvate and phosphoenolpyruvate

The last irreversible reaction of glycolysis is the conversion of phosphoenolpyruvate and ADP to pyruvate and ATP. The reaction has a large negative free energy change reflecting the fact that PEP has a higher phosphorylation potential than ATP, and the mass action ratio in the cell in liver, as in muscle, is a long way from equilibrium. In contrast to the situation in muscle, pyruvate kinase is an important site of regulation in

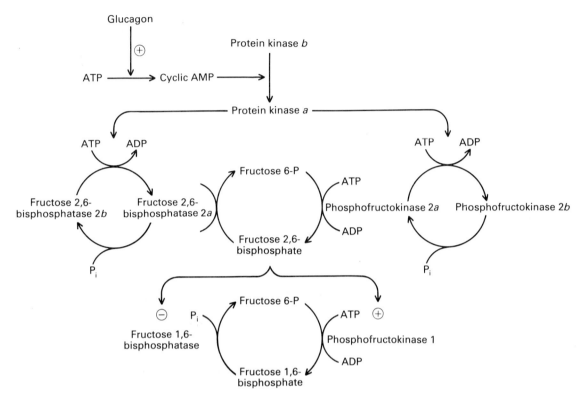

Fig. 15.8. The gluconeogenic cascade. Phosphorylation by protein kinase, in response to elevated cyclic AMP, activates fructose 2,6-bisphosphatase and inhibits phosphofructokinase 2 leading to a fall in fructose 2,6-bisphosphate. This increases fructose 1,6-bisphosphate activity and decreases phosphofructokinase 1 activity, thereby activating gluconeogenesis.

liver. It requires the energy from two molecules of ATP to reverse the pyruvate kinase reaction. This is achieved by the sequence of reactions shown in Fig. 15.9. The key enzymes are pyruvate carboxylase, which converts pyruvate to oxaloacetate and is exclusively mitochondrial, and phosphoenolpyruvate carboxykinase (PEP-CK), which converts oxaloacetate to phosphenolpyruvate. Depending upon the species, PEP-CK may be predominantly cytoplasmic, predominantly mitochondrial or more or less evenly distributed between both compartments. Oxaloacetate is always formed from pyruvate in the mitochondria. If PEP-CK is cytoplasmic, there is a requirement for transfer of oxaloacetate from the mitochondria to the cytoplasm. This requires a shuttle mechanism since the mitochondrial inner membrane is impermeable to oxaloacetate. The nature of the shuttle depends upon the gluconeogenic substrate.

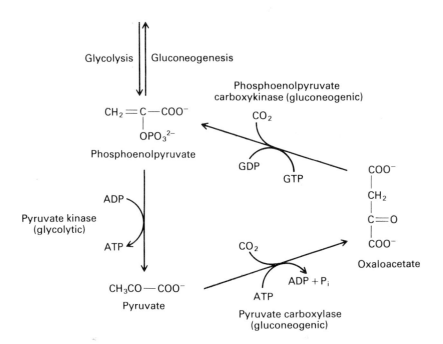

Fig. 15.9. The pyruvate/phosphoenolpyruvate substrate cycle.

1 If pyruvate is the substrate, oxaloacetate is reduced to malate which leaves the mitochondria to be reoxidized to oxaloacetate (Fig. 15.10a). This has the effect of transferring reducing power as NADH from the mitochondria to the cytoplasm. NADH will be needed to drive glyceraldehyde 3-phosphate dehydrogenase in the gluconeogenic direction, since there is no other source of NADH in the cytoplasm.

2 If lactate is the substrate, the oxidation of lactate to pyruvate provides a source of cytoplasmic NADH. In this case a different shuttle for oxaloacetate is needed which does not transfer reducing power. This can be achieved using aspartate, glutamate and α-oxoglutarate (Fig. 15.10b).

3 If PEP-CK is mitochondrial, there is no need for oxaloacetate transfer, PEP formed in the mitochondria is transferred to the cytoplasm (Fig. 15.10c). If the original substrate was lactate, NADH will be available for the reversal of glyceraldehyde 3-phosphate dehydrogenase. If the substrate was pyruvate, however, a mechanism for the transfer of reducing equivalents from the mitochondrion to the cytoplasm is necessary and it is not clear how this occurs.

4 If PEP-CK is present in both the mitochondria and the cytoplasm the situation is simple. When pyruvate is the substrate, the pathway shown in Fig. 15.10a will operate; if lactate is the substrate, the pathway shown in Fig. 15.10c will operate.

263 Regulation of metabolism in liver

(a)
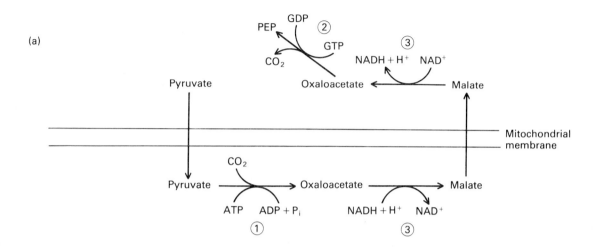

(b)
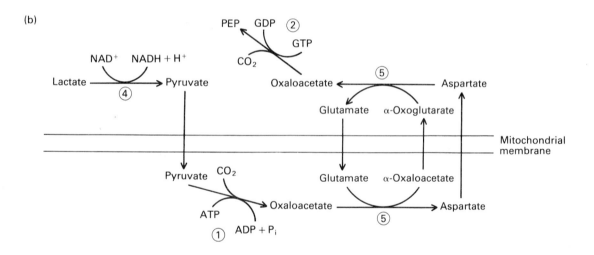

Fig. 15.10. Gluconeogenic shuttles for the generation of cytoplasmic PEP from mitochondrial oxaloacetate. (a) PEP-CK exclusively cytoplasmic (pyruvate as substrate). Oxaloacetate is transferred out of the mitochondria by a malate shuttle which also provides the NADH for the reversal of glyceraldehyde 3-P dehydrogenase. (b) PEP-CK exclusively cytoplasmic (lactate as substrate). In this case, cytoplasmic reducing power is generated by the oxidation of lactate to pyruvate, so oxaloacetate must be transferred to the cytoplasm with no transfer of reducing power. This is achieved by an aspartate transamination shuttle. (c) PEP-CK mitochondrial (lactate as substrate). PEP is transported directly into the cytoplasm and cytoplasmic reducing power is supplied by lactate oxidation. (d) Connection with urea cycle when alanine is substrate. Cytoplasmic reducing power is supplied by the oxidation of malate to

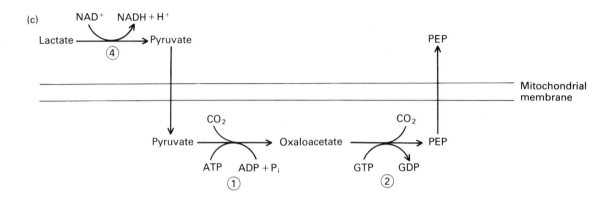

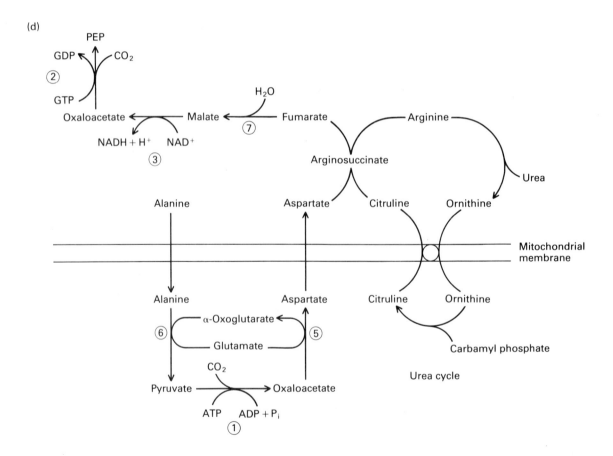

oxaloacetate. Aspartate may feed into gluconeogenesis in the cytoplasm via the urea cycle. The reactions involved are: (1) pyruvate carboxylase, (2) phosphoenolpyruvate carboxykinase (PEP-CK), (3) malate dehydrogenase, (4) lactate dehydrogenase, (5) aspartate aminotransferase, (6) alanine aminotransferase, (7) fumarase.

5   Amino acids such as alanine and aspartate may act as a source of oxaloacetate through a connection with the urea cycle (Fig. 15.10d).

The enzymes common to all these pathways are pyruvate carboxylase, which converts pyruvate to oxaloacetate, and PEP-CK, which converts oxaloacetate to PEP. These two enzymes constitute the gluconeogenic side of a substrate cycle, of which the glycolytic side is pyruvate kinase (Fig. 15.9). Under gluconeogenic conditions pyruvate kinase might be expected to tbe inhibited and either pyruvate carboxylase or PEP-CK, or both, activated.

15E4 Regulation of pyruvate/oxaloacetate/PEP substrate cycle

Pyruvate kinase in liver is highly regulated. The reaction rate has a sigmoid relationship to the concentration of the substrate PEP indicating cooperativity (see Section 1F2), and there are a number of allosteric effectors. The K_m for PEP is increased by alanine, phenylalanine and ATP, and decreased by fructose 1,6-bisphosphate. During gluconeogenesis many of these effectors will change so as to decrease the activity of pyruvate kinase. Alanine and phenylalanine are gluconeogenic substrates, and inhibition of phosphofructokinase and activation of fructose bisphosphatase leads to a reduction in the concentration of fructose 1,6-bisphosphate. Liver pyruvate kinase is also phosphorylated by the cyclic AMP-dependent protein kinase. This has the effect of intensifying the inhibition of the enzyme; the K_m for PEP is increased, the K_a for fructose 1,6-bisphosphate is increased, and the K_I for alanine and for ATP is decreased. Thus glucagon, acting via cyclic AMP, inhibits the enzyme. The allosteric effectors also modulate the phosphorylation, alanine making the enzyme a better substrate for phosphorylation, and fructose 1,6-bisphosphate making it a poorer substrate for phosphorylation. Thus the effects of the intrinsic and extrinsic regulators are synergistic. During prolonged starvation the concentration of pyruvate kinase in liver decreases, presumably as a result of the action of glucagon.

It seems likely that the other half of the substrate cycle is controlled at either pyruvate carboxylase or PEP-CK, since these are the two enzymes which are always involved (see Fig. 15.9). The total activity of both enzymes is low. It is not possible to measure the mass action ratio of the two reactions individually mainly because of the difficulty of determining the relevant oxaloacetate concentration. Oxaloacetate is found in both the mitochondria and the cytoplasm, the concentration in both is low and it is very unstable. However, the two enzymes can be considered as a single step described by the equation:

pyruvate + ATP + GTP $\longrightarrow$ PEP + GDP + ADP + P_i.

The mass action ratio for this is 10^4 times lower than the equilibrium constant. This should be treated with caution since no account is taken of the separation of substrates between the mitochondria and the cytoplasm. However, the difference is so large as to make it very likely that the overall process is rate limiting.

Pyruvate carboxylase is activated by acetyl CoA. Thus under gluconeogenic conditions the enzyme will be activated by the high levels of acetyl CoA resulting from fatty acid oxidation (see Section 15B1). This might seem a rather indirect control for a key enzyme but pyruvate carboxylase is not exclusively concerned with gluconeogenesis. It is also in the pathway of fatty acid biosynthesis (see Chapter 16) and plays a general role in maintaining the level of oxaloacetate to allow citric acid cycle function. The elevated level of acetyl CoA and NADH will also inhibit pyruvate dehydrogenase by the same mechanisms as operate in muscle (see Section 14J1). This inhibits the oxidation of pyruvate derived from the circulation, or from lactate, or by transamination from alanine. The control of PDH and pyruvate carboxylase by acetyl CoA will therefore tend to direct pyruvate towards gluconeogenesis rather than to acetyl CoA. This is crucial since in mammals acetyl CoA cannot act as a substrate for gluconeogenesis and once pyruvate in converted to acetyl CoA it is lost as a source of glucose.

PEP-CK seems to be a very likely candidate for a control site. It is the step which connects the carbon flow from gluconeogenic substrates to gluconeogenesis and can be regarded as the first exclusively gluconeogenic enzyme. However, the free energy change of the reaction is small. There is some indication that the reaction is regulated since the concentrations of malate and aspartate fall when gluconeogenesis is increased. Malate and asparate are thought to equilibrate with oxaloacetate through the enzymes, malate dehydrogenase and aspartate aminotransferase (Fig. 15.5). It seems, therefore, that the concentration of oxaloacetate, the substrate for PEP-CK, falls on activation of gluconeogenesis indicating activation of PEP-CK. A great deal of effort has been expended attempting to find allosteric effectors of PEP-CK without success. Furthermore, the enzyme is monomeric, which argues against its being subject to allosteric regulation (see Section 1F3). In starvation the concentration of PEP-CK increases in response to glucagon. This effect is mediated by cyclic AMP and requires the presence of glucocorticoids. Insulin or high levels of glucose prevent the increase in activity. It seems that the main control on PEP-CK is by alteration of its concentration in the cell. Consistent with this is the observation that PEP-CK has one of the highest known rates of turnover of any enzyme ((see Section 1H2), and changes in concentration in a few hours. There is normally no need to activate gluconeogenesis rapidly since short-term changes in glucose requirement can be supplied by glycogen breakdown. Enzyme induction is therefore a perfectly adequate control mechanism.

15F Conclusions

The pathways of glycolysis and gluconeogenesis and the regulatory mechanisms operating in liver are shown in Fig. 15.7. Regulation operates at several different levels. At the level of glucose entry and phosphorylation, control is exercised directly by the blood glucose concentration. At the fructose 6-phosphate/fructose 1,6-bisphosphate substrate cycle, control is exercised by a cascade mechanism, glucagon operating through cyclic AMP-dependent protein kinase, and changes in the concentration of fructose 2,6-bisphosphate.

Many controls operate at the level of the pyruvate/PEP substrate cycle. In the glycolytic direction, pyruvate kinase is controlled allosterically and by cyclic AMP-dependent phosphorylation. The control of pyruvate carboxylase and pyruvate dehydrogenase to promote oxaloacetate rather than acetyl CoA synthesis, is largely through the increase in the concentration of acetyl CoA arising from increased fatty acid oxidation. This in turn is an indirect consequence of the activation of lipolysis in adipose tissue by hormones. PEP-CK is controlled by changing its concentration. Thus gluconeogenesis is activated by a combination of alterations in the metabolic state and direct hormone effects.

Chapter 16   Regulation of metabolism in adipose tissue

16A   Introduction

Mammals have two types of adipose tissue: white fat, which is concerned with the storage and release of lipid as a fuel reserve, and brown fat, which is used to generate heat. Brown fat contains many more mitochondria than white fat and this is what gives it its colour. Stored lipid is oxidized in the tissue itself and the mitochondria can be uncoupled in a controlled fashion so that the oxidation generates heat. Brown fat is particularly important in the new-born and in hibernating mammals. Recently it has been suggested that uncoupled oxidation in brown fat serves as a means to oxidize food which is in excess of requirements and would otherwise tend to be stored as lipid, leading to excessive weight gain.

The most extensively studied source of white fat is the epididymal fat pad of the male rat. Fat cells can be isolated by collagenase digestion yielding a homogeneous preparation with no connective tissue cells. Rat fat cells synthesize fatty acids from glucose and take in fatty acids from circulating lipoproteins. Both are esterified to form triacylglycerol which is stored. During fasting or exercise, triacylglycerol is hydrolysed and fatty acids and glycerol are released. Rat fat cells respond to a large number of different hormones (Table 16.1). Since the metabolism is relatively simple, this has made them very attractive as a system in which to study mechanisms of action of hormones. In particular, they have been widely used in studies on the mechanism of action of insulin.

Table 16.1. Hormones which have acute effects on rat fat cell metabolism

Insulin
Adrenaline α_1
Adrenaline β
ACTH
Glucagon
Secretin
TSH (thyroid stimulating hormone)
Prostaglandin E_1
Adenosine

16B Fatty acid biosynthesis

The pathway of fatty acid biosynthesis from glucose is shown in Fig. 16.1. Glucose is converted to mitochondrial acetyl CoA via glycolysis and pyruvate dehydrogenase. Acetyl CoA leaves the mitochondria by a citrate shuttle and is converted to fatty acids in the cytoplasm. In fat cells, almost all the glucose taken up is used either as a carbon source for triacylglycerol biosynthesis or to provide the energy and reducing power needed for triacylglycerol synthesis. Other substrates which can be used for fatty acid synthesis include acetate, pyruvate and fructose. Acetate is converted to acetyl CoA in the cytoplasm by the enzyme acetyl CoA synthetase which catalyses the reaction:

acetate + CoASH + ATP $\rightleftharpoons$ acetyl CoA + AMP + PP_i.

Pyruvate enters the pathway at the level of pyruvate dehydrogenase. Fructose is metabolized via essentially the same pathway as glucose. Fructose is probably not an important source of carbon for fatty acid biosynthesis in fat cells *in vivo* since most dietary fructose is taken up and metabolized by the liver.

16B1 Cytoplasmic reactions of fatty acid biosynthesis

Cytoplasmic acetyl CoA is carboxylated to malonyl CoA by the enzyme acetyl CoA carboxylase (Fig. 16.2). This is an activation step and the CO_2 moiety is removed during fatty acid biosynthesis. Thus CO_2 is essential for fatty acid biosynthesis but is not itself incorporated into fatty acids.

Fatty acid synthetase catalyses a total of seven reactions. In bacteria such as *Escherichia coli*, the seven catalytic activities and an acyl carrier protein are on individual separable protein subunits. In mammals, a single multi-functional enzyme appears to catalyse all seven reactions. The enzyme contains a 4'-phosphopantetheine group, which is identical to the functional group of coenzyme A and acts as a carrier of acyl groups. The sequence of reactions catalysed by fatty acid synthetase is shown in Fig. 16.3. Synthesis stops when the chain length reaches sixteen carbon atoms and palmitate is released. Longer chain fatty acids, such as stearate with eighteen carbons, are made by essentially the same series of reactions using palmitoyl CoA and malonyl CoA as starting materials. A different enzyme is involved which is associated with the endoplasmic reticulum.

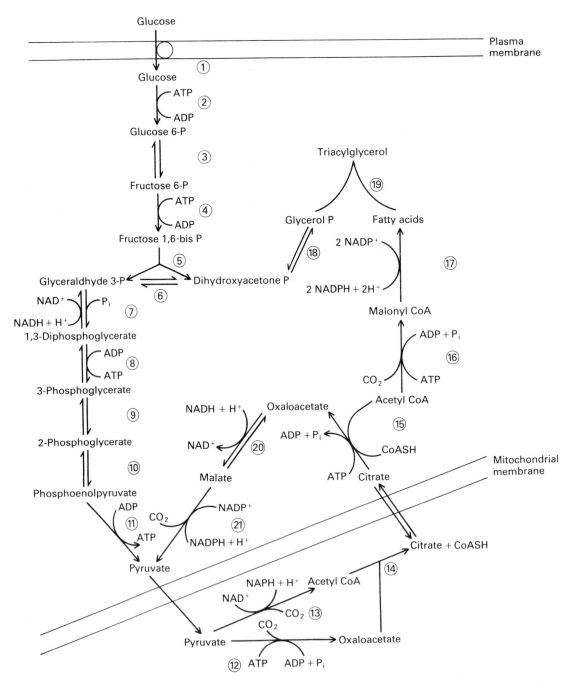

Fig. 16.1. Synthesis of fatty acids and triacylglycerol from glucose in fat cells. The reactions involved are: (1) glucose entry, (2) hexokinase, (3) glucose 6-P isomerase, (4) phosphofructokinase, (5) aldolase, (6) triose phosphate isomerase,

(7) glyceraldehyde 3-P dehydrogenase, (8) phosphoglycerate kinase, (9) phosphoglycermutase, (10) enolase, (11) pyruvate kinase, (12) pyruvate carboxylase, (13) pyruvate dehydrogenase, (14) citrate synthase, (15) ATP-citrate lyase, (16) acetyl CoA carboxylase, (17) fatty acid synthetase, (18) glycerol phosphate dehydrogenase, (19) esterification, (20) NAD-malate dehydrogenase, (21) NADP-malate dehydrogenase (malic enzyme).

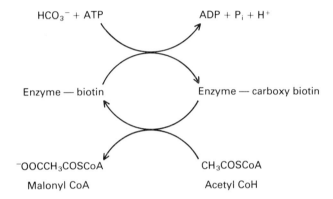

Fig. 16.2. The acetyl CoA carboxylase reaction. The enzyme is first carboxylated on the biotin prosthetic group and CO_2 is then transferred to acetyl CoA.

16B2 Transfer of acetyl units across the mitochondrial membrane

Unless acetate itself is the carbon source, acetyl CoA is produced in the mitochondria via the pyruvate dehydrogenase reaction. Since fatty acid synthesis takes place in the cytoplasm and the mitochondrial inner membrane is impermeable to acetyl CoA, a shuttle mechanism is required. This is provided by the pyruvate malate cycle (Fig. 16.4). Citrate is formed in the mitochondria from acetyl CoA and oxaloacetate, by citrate synthase, and leaves the mitochondria on the citrate carrier. Acetyl CoA and oxaloacetate are regenerated in the cytoplasm by the action of ATP-citrate lyase (citrate cleavage enzyme). This completes the transfer of acetyl CoA to the cytoplasm, but, in order to complete the cycle, oxaloacetate must be returned to the mitochondria, and the mitochondrial inner membrane is impermeable to oxaloacetate. Cytoplasmic malate dehydrogenase catalyses the NADH-linked reduction of oxaloacetate to malate which is then oxidatively decarboxylated to pyruvate by NADP-malate dehydrogenase (malic enzyme) reducing NADP to NADPH (Fig. 16.4). The effect of this part of the cycle is to transfer reducing power from cytoplasmic NADH to cytoplasmic NADPH. The cycle is completed by the entry of pyruvate into the mitochondria and its carboxylation to oxaloacetate.

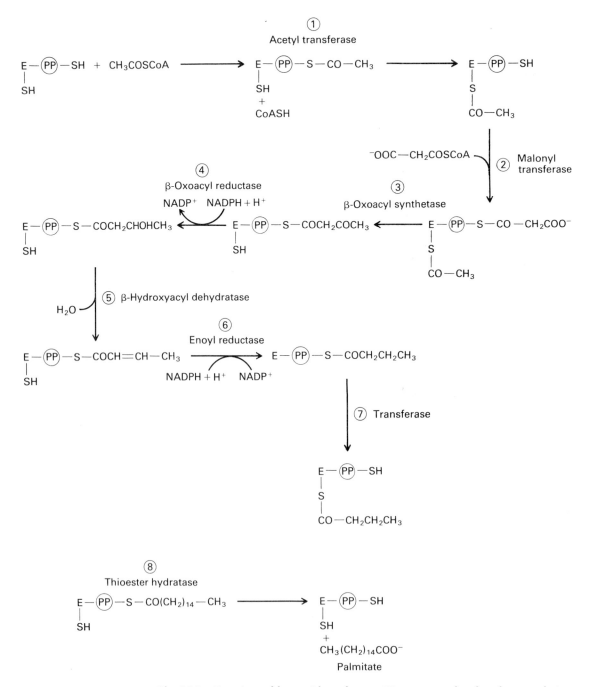

Fig. 16.3. Reactions of fatty acid synthetase. PP represents the phosphopantetheine prosthetic group. The process is repeated 7 times to yield a chain length of 16 (palmitate) when the fatty acid is hydrolysed of the phosphopantetheine rather than being transferred.

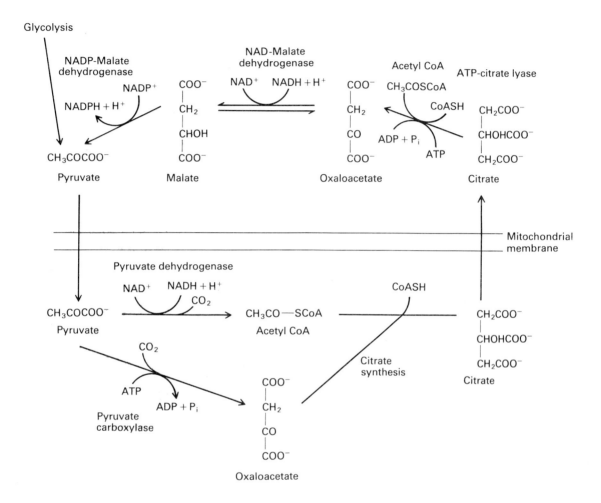

Fig. 16.4. The pyruvate/malate cycle. For a full turn of the cycle, two molecules of pyruvate are involved—one to generate acetyl CoA and one to generate oxaloacetate. The cycle transfers one acetyl CoA from the mitochondria to the cytoplasm and forms one cytoplasmic NADPH from NADH, at the expense of two ATP.

For the expenditure of two molecules of ATP, the pyruvate/malate cycle transfers one acetyl CoA from the mitochondria to the cytoplasm and generates one NADPH from NADH in the cytoplasm. The source of NADH is glycolysis. The glyceraldehyde 3-phosphate dehydrogenase reaction produces one NADH for every acetyl CoA. Two molecules of NADPH are required to incorporate one acetyl unit into fatty acid, so the pyruvate/malate cycle might be expected to provide exactly 50% of the NADPH needed, the rest being provided by the pentose phosphate pathway (see Section 16B3). It is quite difficult to measure the relative contributions of the two pathways, but it seems likely that the pyruvate/malate cycle accounts for only 40% of the NADPH requirement under optimal conditions for fatty acid biosynthesis. If this is the case, about 20% of the flow must bypass the NADP-malate dehydrogenase step by the entry of malate into the mitochondria (Fig. 16.5). This has the effect of transferring some of the reducing power from glycolysis into the mitochondria.

16B3 The pentose phosphate pathway

The pentose phosphate pathway provides an alternative pathway to

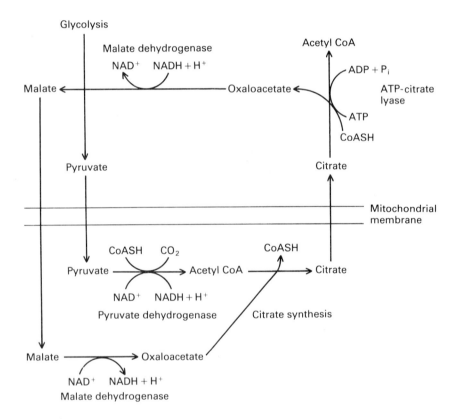

Fig. 16.5. Alternative pathway for the pyruvate/malate cycle. In this case there is no transfer of cytoplasmic reducing power from NADH to NADP-malate dehydrogenase and regenerating oxaloacetate from pyruvate via pyruvate carboxylase. Instead, oxaloacetate is regenerated in the mitochondria from malate, leading to a transfer of reducing power from the mitochondria to the cytoplasm.

glycolysis for the oxidation of glucose. The oxidative part of the pathway is catalysed by the first two enzymes (Fig. 16.6). Glucose 6-phosphate dehydrogenase catalyses the oxidation of glucose 6-phosphate to 6-phosphogluconolactone and the subsequent hydration of the lactone to 6-phosphogluconate. 6-Phosphogluconate dehydrogenase catalyses the oxidative decarboxylation of 6-phosphogluconate to ribulose 5-phosphate. Both reactions reduce NADP to NADPH and both are irreversible. If the pentose phosphate pathway provides 50% of the NADPH for fatty acid biosynthesis, the rate of flow through these two enzymes will be equal to the rate of flow to acetyl CoA through glycolysis. One glucose produces two molecules of acetyl CoA, each of which needs two molecules of NADPH to be incorporated into fatty acids. Therefore, a total of four molecules of NADPH is needed. Two of these can be supplied by the pyruvate/malate cycle, one for each acetyl unit. The other two come from the pentose phosphate pathway, so for each glucose converted to acetyl CoA another must enter the pentose phosphate pathway to generate NADPH. In fact the rate of flow in this direction will be

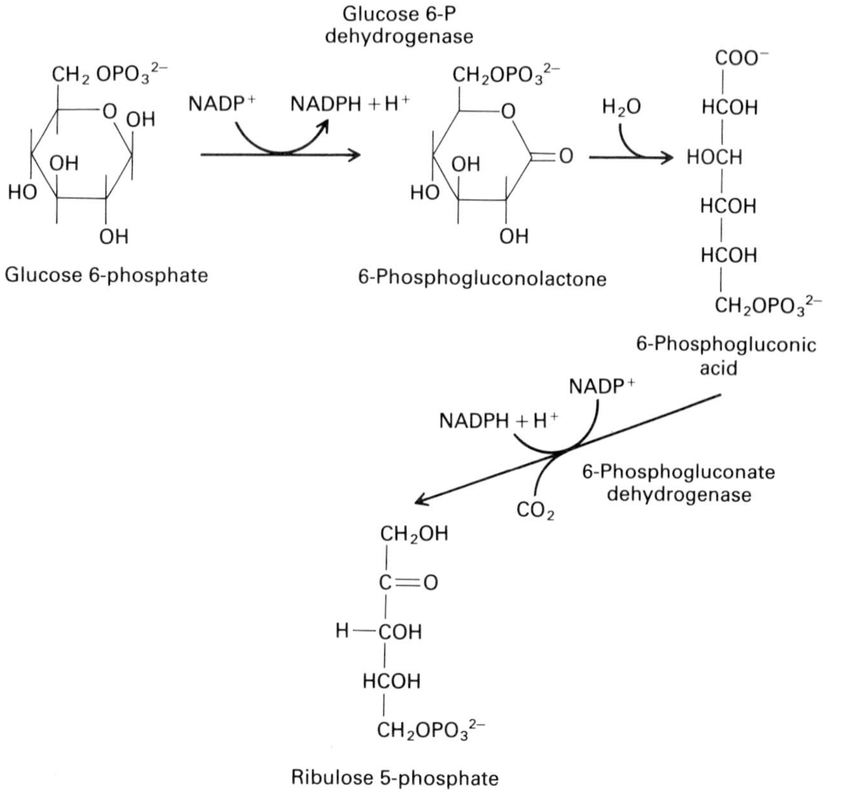

Fig. 16.6. The oxidative pentose phosphate pathway.

slightly greater since the pentose phosphate pathway provides rather more than 50% of NADPH.

Substantial amounts of ribulose 5-phosphate are formed. This is diverted back to the glycolytic pathway by a series of interconversions between sugar phosphates. Ribulose 5-phosphate may be converted to xylulose 5-phosphate by an epimerase, or ribose 5-phosphate by an isomerase. These can yield fructose 6-phosphate and glyceraldehyde 3-phosphate by a series of two- and three-carbon transfers catalysed by the enzymes transaldolase and transketolase (Fig. 16.7). The overall stoichiometry of the pathway can be represented by the equation:

6 NADP + 3 glucose 6-P $\longrightarrow$ 2 fructose 6-P + glyceraldehyde 3-P + 3 CO_2 + 6 NAPDH + 6 H^+.

High activities of the two dehydrogenases are found in tissue which require large amounts of NADPH to provide reducing power for synthesis. The non-oxidative reactions occur in all tissues and are freely reversible. They can serve to generate ribose 5-phosphate for nucleotide biosynthesis from fructose 6-phosphate and glyceraldehyde 3-phosphate, as well as to convert ribulose 5-phosphate to glycolytic intermediates. The direction of flow in the non-oxidative pathway will depend upon the relative requirements for NADPH and ribose 5-phosphate. In adipose tissue, far more NADPH than ribose 5-P is required.

16B4 Energy balance in fatty acid biosynthesis: control of the citric acid cycle

The conversion of glucose to fatty acids leads to net production of ATP. From Fig. 16.8 and Table 16.2 it can be seen that, when the pathway uses the pyruvate/malate cycle, there is a net production of one molecule of ATP for each acetyl unit incorporated. When all the NADPH required is generated from the pentose phosphate pathway, six molecules of ATP are produced for each acetyl unit incorporated. It has been suggested that the maximum rate of fatty acid biosynthesis in rat adipose tissue is limited by the capacity of the cell to make use of the ATP which is generated. There is some experimental support for this since the maximum rate measured, in the presence of insulin and glucose, is close to the expected rate calculated on the basis of the measured ATP requirements of the cell. Also, if small amounts of uncoupler are added, so as to increase the rate of ATP breakdown without causing any marked reduction in the ATP concentration, then fatty acid synthesis is activated rather than inhibited. Acetate and glucose together support a higher rate of fatty acid synthesis than the sum of the rate with glucose alone and

Regulation of metabolism in adipose tissue

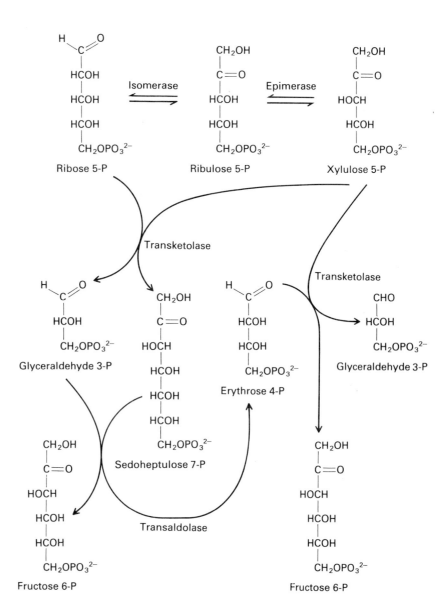

Fig. 16.7. Non-oxidative pentose phosphate pathway.

the rate with acetate alone. This is thought to be a result of net ATP utilization for the synthesis of acetyl CoA from acetate leading to a reduction in the restriction of synthesis from glucose.

There is no requirement for the citrate cycle to operate to produce energy for fatty acid synthesis from glucose, and measured rates of flow in the citrate cycle beyond citrate synthase are very low. In rat fat cells the reason for this is simple: the total activity of aconitase in the

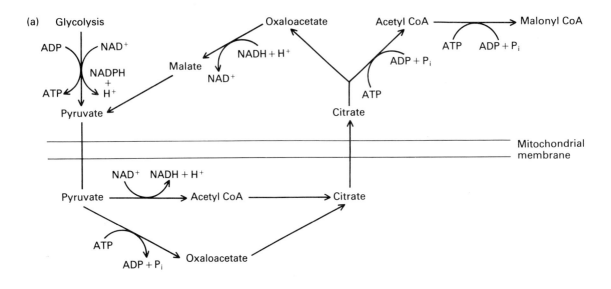

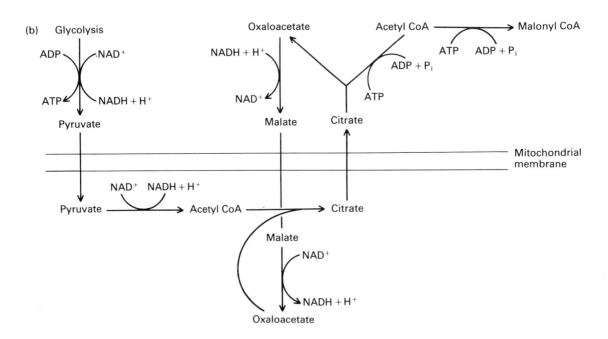

Fig. 16.8. ATP and NADH production and utilization in the conversion of glucose to fatty acids. One acetyl unit derived from glucose is incorporated into fatty acid. Other cofactors involved are omitted. (a) Pyruvate/malate cycle operating. One ATP is

produced and three utilized, making a net utilization of two. Two molecules of NADH are produced and one is utilized making a net production of one NADH. One NADH is equivalent to three ATP so there is a net production of one ATP. (b) Pyruvate/malate cycle (alternative pathway operating). If the pyruvate/malate cycle is short-circuited by the entry of malate into the mitochondria, one ATP is produced and two are utilized making a net utilization of one. Three molecules of NADH are produced and one utilized making a net production of two, equivalent to six ATP. Overall there is a net production of five ATP. The calculation is described in more detail in Table 16.2.

Table 16.2. Energy balance in fatty acid biosynthesis

	ATP equivalents/acetyl CoA incorporated	
	ATP	NADH
Pyruvate/malate cycle operating		
Glycolysis	+1	+3
Pyruvate carboxylase	−1	
Pyruvate dehydrogenase		+3
ATP-citrate lyase	−1	
Acetyl CoA carboxylase	−1	
Malate dehydrogenase		−3
Oxaloacetate to malate in cytoplasm		
Total	−2	+3
Total: 1 ATP excess		
Pyruvate/malate cycle not operating		
Glycolysis	+1	+3
Pyruvate dehydrogenase		+3
ATP-citrate lyase	−1	
Acetyl CoA carboxylase	−1	
Malate dehydrogenase		
Oxaloacetate to malate in cytoplasm		−3
Malate to oxaloacetate in mitochondria		+3
Total	−1	+6
Total: 5 ATP excess		

mitochondria is very low so that citrate tends to be exported and reconverted to oxaloacetate and acetyl CoA rather than converted to isocitrate (Fig. 16.4). Control at the other likely regulatory sites, isocitrate dehydrogenase and α-oxoglutarate dehydrogenase, then becomes irrelevant since there is little carbon flow beyond citrate.

A mechanism is also rquired to restrict the oxidation of malate to oxaloacetate since the entry and oxidation of malate short-circuits the pyruvate/malate cycle (Fig. 16.8b). This is a less efficient pathway since it generates no NADPH and it transfers reducing power into the mitochondria as NADH, leading to ATP synthesis which might restrict the rate of fatty acid biosynthesis (Fig. 16.8, Table 16.2). It is not clear how malate oxidation is restricted. The activity of malate dehydrogenase is high in both the mitochondria and the cytoplasm, and the reaction is likely to be close to equilibrium and not rate limiting. It may be subject to product inhibition by oxaloacetate generated through the pyruvate carboxylase reaction. It has been suggested that the maximum rate of the pyruvate/malate cycle is governed by the V_{max} of pyruvate carboxylase. Thus, as the rate of fatty acid synthesis increases, a point may be reached when pyruvate carboxylase is operating at maximum capacity. Further increases in the rate of fatty acid biosynthesis might lead to a fall in oxaloacetate and activation of the conversion of malate to oxaloacetate, short-circuiting the pyruvate/malate cycle. This is an attractive hypothesis, but there is no direct evidence in its favour. An alternative might be regulation of the transport of malate into the mitochondria, but there is no evidence to support this either.

Both pyruvate and acetate can be used as substrates for fatty acid biosynthesis, although maximum rates are considerably lower than with glucose. Obviously, neither glycolysis nor the pentose phosphate pathway will function so that there is no obvious source of cytoplasmic reducing power. The citrate cycle then functions as a source of reducing power. The likely pathway is shown in Fig. 16.9. Cytoplasmic NADPH is generated by running the citrate cycle from citrate to α-oxoglutarate, and from malate via pyruvate to oxaloacetate, in the cytoplasm.

Under lipolytic conditions the ATP requirements of fat cells are likely to be satisfied by β-oxidation of some of the endogenous lipid store. The capacity for β-oxidation in fat cells is low, but, in the absence of fatty acid biosynthesis, the ATP requirement will also be very limited. There may be enough mitochondrial citrate cycle capacity to oxidize the amounts of acetyl CoA involved. If not, the mitochondrial aconitase step can be bypassed by using the enzyme in the cytoplasm.

Regulation of metabolism in adipose tissue

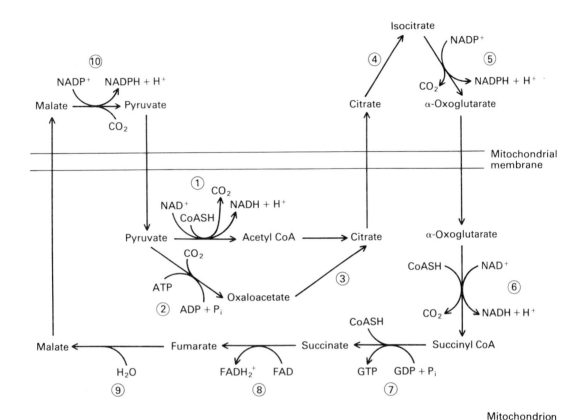

Fig. 16.9. Generation of cytoplasmic reducing power by the citrate cycle. The citrate cycle from citrate to α-oxoglutarate runs in the cytoplasm using cytoplasmic aconitase and cytoplasmic isocitrate dehydrogenase which produces NADPH. The mitochondrial malate dehydrogenase step is bypassed by the export of malate to the cytoplasm where it is converted to pyruvate yielding NADPH. Pyruvate enters the mitochondrion and is converted to oxaloacetate completing the citrate cycle. Other citrate cycle reactions are as usual. The reactions involved are: (1) pyruvate dehydrogenase, (2) pyruvate carboxylase, (3) citrate synthase, (4) aconitase, (5) $NADP^+$-isocitrate dehydrogenase, (6) α-oxoglutarate dehydrogenase, (7) succinate thiokinase, (8) succinate dehydrogenase, (9) fumarase, (10) NADP-malate dehydrogenase (malic enzyme).

16C Short-term regulation of fatty acid biosynthesis

Fatty acid synthesis is regulated by insulin, which activates the process, and adrenaline or glucagon, which inhibit it. There are both acute effects of hormones and longer-term changes in response to changes in the diet, which are also mediated by hormones. In fat cells, insulin has short-term effects at three sites in the pathway from extracellular glucose to fatty acids.

16C1 Glucose transport

Glucose transport in fat cells is strongly stimulated by insulin and current ideas about the mechanism of action are discussed in Chapter 10. It is probable that, in the absence of insulin, fatty acid biosynthesis is limited by the rate of entry of glucose into the cell. Fifteen or twenty years ago this was thought to be the only site of action of insulin, but it became apparent that there must be at least one additional site of regulation and the evidence pointed to a regulation at the level of pyruvate metabolism. Fat cells convert glucose to fatty acids and also release pyruvate and lactate derived from glucose. Insulin stimulates both, but the effect on fatty acid biosynthesis may be as high as twentyfold while the effect on lactate and pyruvate release is only about twofold.

Fructose can also be used as a substrate for fatty acid biosynthesis, and for pyruvate and lactate production. In the absence of insulin, fructose uptake is considerably faster than glucose uptake. Insulin only increases the rate of fructose uptake by about 50%. However, insulin doubles the rate of fatty acid biosynthesis, while the rate of release of lactate and pyruvate falls by 40%.

Fat cells contain significant amounts of glycogen which can be used as a substrate for fatty acid biosynthesis and lactate and pyruvate output in the absence of extracellular glucose. Insulin increases fatty acid biosynthesis from glycogen threefold, while lactate and pyruvate output is reduced by 70% (see Table 16.3). This also raises the interesting possibility that, in the absence of glucose, insulin activates, rather than inhibits, glycogen breakdown.

Finally, fatty acid biosynthesis from pyruvate itself is activated about twofold by insulin. It seems likely that a fall in pyruvate release reflects a fall in the cellular concentration of pyruvate. This would be consistent with the activation of one of the enzymes metabolizing pyruvate – either pyruvate dehydrogenase (PDH) or pyruvate carboxylase (see Fig. 16.1).

Table 16.3. Effect of insulin on fatty acid synthesis and lactate and pyruvate production from various substrates

Substrate	Rate as % control	
	Lactate + pyruvate	Fatty acids
Glucose	200	1500
Fructose	60	180
Pyruvate	–	162
Endogenous glycogen	33	144

16C2 Pyruvate dehydrogenase

Adipose tissue PDH is subject to the same regulation as the enzyme in muscle (see Section 14J1 and Fig. 16.10). The effect of insulin results from an increase in the proportion of the enzyme in the dephosphorylated state. The effect is antagonized by adrenaline, but, in isolated fat cells, adrenaline does not reduce basal levels of PDH in the absence of insulin. The nature of the messenger for the insulin effect is still not established. Insulin does not change or cause a slight increase in the ATP/ADP, acetyl CoA/CoA and NADH/NAD ratios depending upon the conditions. An increase in these ratios would be expected to increase phosphorylation and inhibit PDH, so none of them can act as the signal for the insulin effect (see Section 14J1 and Fig. 16.10). Insulin may either increase or reduce pyruvate levels (see above) but always activates PDH so a change in pyruvate cannot be the signal. Of the known effectors, this leaves Ca^{2+} which activates the phosphatase and inhibits the kinase leading to reduced phosphorylation and activation of the enzyme.

The evidence in favour of Ca^{2+} as a second messenger for insulin is discussed in detail in Chapter 10. There is quite an impressive body of circumstantial evidence which implies that insulin changes cytoplasmic Ca^{2+} levels in fat cells, but direct measurements are difficult. The activation of PDH, an enzyme known to be sensitive to Ca^{2+}, has itself been used as an argument in favour of Ca^{2+} as an insulin second messenger. The proposed peptide second messenger for insulin has been reported to activate PDH in isolated mitochondria, but again the nature of the direct effect on the enzyme is not known.

The effect of insulin on PDH activity is quite small; typically the activity doubles. However, the changes in metabolite concentration on

Fig. 16.10. Regulation of pyruvate dehydrogenase by phosphorylation.

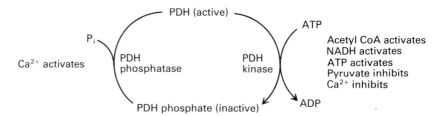

the addition of insulin might be expected to cause a reduction in PDH activity. In the absence of insulin, PDH activity is much greater than the rate of fatty acid biosynthesis from glucose, and the reaction is not likely to be rate limiting. In the presence of insulin, PDH activity is much closer to the rate of fatty acid synthesis from glucose. However, when the substrate is glucose, insulin probably causes an increase rather than a fall in the concentration of the substrate pyruvate, suggesting that PDH is not a rate-controlling step for fatty acid synthesis. If fructose or endogenous glycogen are substrates, insulin causes a reduction in pyruvate release which presumably reflects a fall in intracellular pyruvate. This suggests that PDH is the rate-controlling step for fatty acid synthesis from these substrates.

16C3 Glycolysis

In fat cells, the role of glycolysis is to provide acetyl CoA for synthesis rather than to respond to the energy demands of the cell. The regulation of glycolysis in fat might therefore be expected to be different from the regulation in muscle. The main factor controlling the rate of glycolysis in fat is the activation of glucose entry by insulin, and there seems to be little regulation at hexokinase or phosphofructokinase. However, the regulatory properties of these enzymes are the same as those of the muscle enzymes. It has been suggested that in fat cells the AMP concentration is always sufficiently high to maintain phosphofructokinase activity and prevent it from becoming rate limiting. The recently discovered fructose 2,6-bisphosphate mechanism, whereby hormones control liver PFK activity directly, would also allow for regulation of fat cell PFK. However, the occurrence of this mechanism in fat cells has not been established.

16C4 Acetyl CoA carboxylase

Insulin activates fatty acid biosynthesis from glucose in liver, but neither glucose transport nor PDH activity are affected. This led to the identification of acetyl CoA carboxylase as a site of regulation. The activity of the enzyme is increased by insulin in both liver and adipose tissue, and inhibited by adrenaline. The regulation of the enzyme is unusual in that activation requires polymerization of inactive monomers, the association of ten or more being required for full activity. Polymerization requires citrate, and depolymerization is favoured by ATP. Neither of these effects are thought to be involved in the regulation of the enzyme in the cell, since changes in the rate of fatty acid biosynthesis do not correlate with changes in the cellular level of either citrate or ATP. Dissociation and inhibition of the enzyme are favoured by micromolar levels of fatty acyl CoA. This may well be physiologically important.

In adipose tissue, insulin activates fatty acid esterification and inhibits lipolysis tending to reduce the fatty acyl CoA concentration, while adrenaline has the reverse effect. In liver, under fasting conditions, there are high levels of fatty acyl CoA in the cytoplasm derived from circulating fatty acids, while in the well-fed state, levels of circulating free fatty acids are lower. Thus, in both tissues, insulin may activate acetyl CoA carboxylase as a result of a reduction in fatty acyl CoA, and adrenaline or glucagon may inhibit as a result of a rise in fatty acyl CoA. In this case fatty acyl CoA is acting as a signal for an indirect effect of the hormones. There are also direct effects: adrenaline or glucagon promote phosphorylation of acetyl CoA carboxylase via the cyclic AMP-dependent protein kinase which leads to inhibition of the enzyme; insulin promotes phosphorylation at a different site on acetyl CoA carboxylase leading to activation. The mechanism of the insulin effect is not known (see Chapter 10).

16C5 ATP-Citrate lyase

Recently there have been a number of studies of phosphorylation of ATP-citrate lyase. The enzyme is phosphorylated by cyclic AMP-dependent protein kinase and, on a different site, by a cyclic AMP-independent protein kinase. Phosphorylation by the cyclic AMP-dependent protein kinase potentiates phosphorylation by the cyclic AMP-independent protein kinase. These observations are very interesting, and it is difficult to think of any reason for such specific effects other than the regulation of the enzyme. However, so far it has not been possible to correlate these phosphorylation changes with physiologically relevant changes in enzyme activity.

16C6 Regulation of NADPH supply

The production of NADPH by the pyruvate/malate cycle is a consequence of the transfer of acetyl CoA from the mitochondria to the cytoplasm. As the carbon flow to fatty acids from glucose increases, so will the production of NADPH (see Section 16B4). The pentose phosphate pathway is independent of the carbon flow from glucose to fatty acids and, in theory, can be regulated so as to supply any amount of NADPH. When glucose is the only substrate, the pathway supplies about 60% of the NADPH required and the pyruvate/malate cycle supplies 40%. If acetate is added in the presence of glucose, each additional acetyl CoA produced requires two NADPH in order to be incorporated into fatty acids (see Section 16B4). As a result, if acetate is added in the presence of glucose, the pentose phosphate pathway is stimulated, and under these conditions it supplies 70–80% of the total NADPH requirement. The mechanism of regulation of the pentose phosphate pathway appears to be straightforward. Both glucose 6-phosphate dehydrogenase and 6-phosphogluconate dehydrogenase are strongly product-inhibited by NADPH. Thus the rate of the oxidative pentose phosphate pathway is governed by the NADPH concentration, in other words by the NADPH demand.

16D Long-term control of fatty acid biosynthesis

The capacity of adipose tissue for synthesis of fatty acids from glucose is reduced by starvation and restored by refeeding. Artificially induced diabetes also reduces the capacity of rat adipose tissue for fatty acid synthesis. A number of enzymes change in concentration during starvation. There is a reduction in the activity of hexokinase, ATP-citrate lyase, acetyl CoA carboxylase, and fatty acid synthetase on the main pathway, glucose 6-phosphate dehydrogenase, and 6-phosphogluconate dehydrogenase in the pentose phosphate pathway, and NADP malate dehydrogenase in the pyruvate/malate cycle. The circulating concentration of insulin appears to be the key factor controlling these effects, but steroid hormones may also be important. The mechanism is not known.

The capacity of adipose tissue for fatty acid synthesis is very strongly repressed by a high fat content in the diet. If the diet also contains carbohydrate, an increased content of lipid in the diet has little effect on the circulating insulin concentration. Thus the effects of increased dietary lipid cannot be explained by a reduction in the level of circulating insulin, and the mechanism of the response is unknown. It may account for the lack of fatty acid synthesis in humans whose diet, at least in the West, has a high fat content.

16E Lipid uptake in adipose tissue

As well as synthesizing fatty acids from glucose, adipose tissue takes up fatty acids derived from triacylglycerol in lipoprotein complexes. There are two main sources: chylomicrons derived from dietary lipid, and very low-density lipoprotein derived from triacylglycerol synthesized in liver (see Section 13C2). Both are processed to release free fatty acids for absorption by lipoprotein lipase, which in adipose tissue, as in muscle, is located on the surface of the capillaries. Fatty acid is taken into the cell and stored as triacylglycerol. In starvation, adipose tissue lipoprotein lipase decreases in activity in contrast to muscle lipoprotein lipase which increases. Insulin increases lipoprotein lipase activity. These effects result from changes in the amount of enzyme present and are caused by changes in the circulating concentration of insulin, adrenaline and, probably, steroid hormones. In the well-fed state, lipid is directed to adipose tissue for storage and, in starvation, to muscle for energy. The adipose tissue pathway is quantitatively very important. Lipid from the circulation accounts for about 50% of stored lipid in the rat, and almost 100% in humans. The pathway in muscle is probably relatively insignificant as an energy source by comparison with the use of free fatty acids and ketone bodies.

16F Esterification

The substrates for the formation of triacylglycerols are glycerol 3-phosphate and fatty acyl CoA. The sequence of reactions is shown in Fig. 16.11. Fatty acid CoA may be the end-product of endogenous fatty acid biosynthesis, in which case it will be predominantly palmitate, or it may be derived from circulating lipoproteins, in which case it may be one of a number of long-chain fatty acids either fully saturated or partially unsaturated. In either case, fatty acids must first be activated by esterification with coenzyme A catalysed by fatty acyl CoA synthetase (Fig. 16.11). The glycerol phosphate for the process is derived from glycolysis. Unlike liver, fat cells have very low activity of glycerol kinase and cannot therefore generate glycerol phosphate directly from glycerol.

Insulin activates and adrenaline inhibits esterification. However, the mechanisms are poorly understood and it is not clear which of the steps involved is rate limiting. Insulin increases the concentration of glycerol phosphate in adipose tissue, and this has been suggested to be important in the stimulation of esterification. However, there is no evidence that glycerol phosphate levels impose a limitation on esterification, and the rate correlates poorly with changes in the glycerol phosphate level. Insulin increases the activity of glycerol phosphate acyl transferase. This enzyme has been reported to be phosphorylated and inhibited by the cyclic AMP-dependent protein kinase. It is the first enzyme of the esterification process proper and therefore an attractive candidate for the

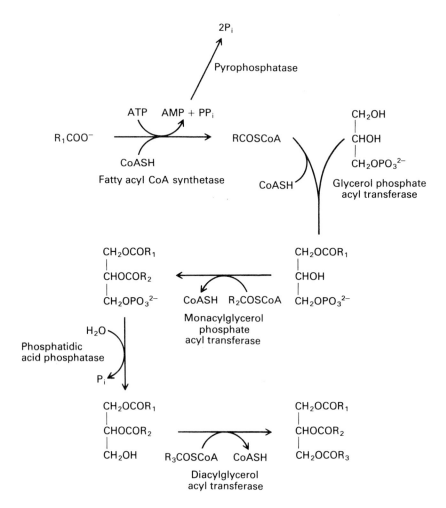

Fig. 16.11. Reactions of triacylglycerol synthesis.

regulatory site. This would provide an explanation for the inhibition of esterification by adrenaline, and in many other instances insulin activates enzymes which are inhibited by cyclic AMP-dependent phosphorylation. The scheme is plausible but, as yet, it has not been shown that the enzyme is rate limiting for the process, and the mechanism of action of insulin is unknown.

16G Lipolysis

Triacylglycerols are broken down to fatty acids and glycerol by three lipases acting sequentially on triacyl-, diacyl- and monoacyl-glycerol

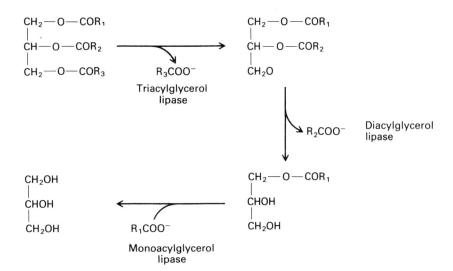

Fig. 16.12. Triacylglycerol breakdown.

(Fig. 16.12). Triacylglycerol lipase is rate limiting for the process, having a much lower activity than the other two enzymes. It is activated by phosphorylation by cyclic AMP-dependent protein kinase. A large number of different hormones activate adenylate cyclase in fat cells (see Chapter 3), but adrenaline and glucagon are the two which are most likely to be physiologically important: adrenaline in stress and exercise, and glucagon in fasting. During starvation, the enzyme increases in concentration and activity, and this depends upon glucocorticoids.

Lipolysis is inhibited by insulin at very low concentrations of the hormone (see Chapter 10). Insulin will lower cyclic AMP levels in fat cells provided that they have been previously elevated by another hormone. However, insulin does not reduce cyclic AMP levels in the absence of other hormones, while it does inhibit lipolysis. Furthermore, under all conditions, the inhibition of lipolysis by insulin correlates very poorly with its effect on cyclic AMP levels. This was discussed in some detail in Section 10G1 when the evidence for the role of changes in cyclic AMP concentration in the action of insulin was evaluated. There are numerous cases where insulin activates an enzyme which is inhibited by cyclic AMP-dependent phosphorylation. In two instances, glycogen synthetase and acetyl CoA carboxylase, this has been shown to result from changes in phosphorylation state at a different site. It is possible that other effects of insulin on enzymes, which antagonize cyclic AMP-mediated effects, may result from phosphorylation at a different site. However, there is no evidence for this in the case of triacylglycerol lipase.

16H Conclusions

The sole function of adipose tissue is to act as a fuel reserve. Regulation of its metabolism is therefore directed towards efficient storage and release of lipid depending upon the available fuel supply and the needs of the animal. It is not surprising, then, that direct hormone effects dominate the regulation of adipose tissue metabolism, since the needs of the whole organism are signalled by the circulating levels of hormone. This contrasts with the situation in muscle where the main regulatory requirement is the need to satisfy its own energy requirements, and as a result the energy status of the cell is the major factor in control mechanisms. Hormones help to determine which fuel supply is used, but not the actual extent of oxidative metabolism.

Index

A23187 97, 120, 128
Acetate
    as carbon source for fatty acid biosynthesis 269, 276–277
Acetoacetate 241, 242–249
Acetylcholine 130
Acetyl CoA 219, 241, 243, 245, 246
    regulation of pyruvate dehydrogenase 238
    in regulation of ketone body synthesis 247
    oxidation in liver 251–254
    in fatty acid biosynthesis 269, 271
Acetyl CoA synthetase 269, 288
Acetyl CoA-acyltransferase 247–249
Acetyl CoA carboxylase
    regulation by Cyclic AMP dependent protein kinase 60
    phosphorylation in liver 124
    insulin dependent phosphorylation 167
    in control of fatty acid biosynthesis 245, 249–251, 269–271, 278–279, 284–286, 289
Acid proteases 26
cis-Aconitate 242
Aconitase 242, 252, 277, 281
ACTH 47
Adenosine 93
    effects on adenylate cyclase 57, 58, 198
Adenylation of proteins 21
Adenylate cyclase 35
    reaction 36–37
    regulation by guanine nucleotides 38–45
    protein components of 45–50
    hormone receptor coupling analysis 47–51
    guanine nucleotide binding proteins 53
    inhibition by hormones 53–57
    non-hormone agonists 57
    non-physiological effectors 57
    effect of calmodulin 57, 113–114, 117, 119
    in platelets 125
    in blow fly salivary gland 128
    in glycogenolytic cascade 24–25, 228
Adenylate kinase 221

Adipose tissue
    physiological function 211
    use of dietary lipid 213
    regulation of metabolism 268–290
ADP 243
    effects on platelet release reaction 123–124
    activation of PDH-kinase 238
ADP-ribosylation 21, 42–43, 46
    poly-ADP-ribosylation 197
Adrenaline
    activation of adenylate cyclase 44, 47
    receptors 91–94
    structure 92
    activation of lipolysis 158–159, 245
    activation of glycogenolysis 228
    mobilization of glucose and lipid 240
    effect on acetyl CoA carboxylase 249, 284–285
    effect on triacylglycerol esterification 287–288
α-Adrenergic receptors 91–94
α_1-Adrenergic receptors
    activation of glucose release 91
    effect on cytoplasmic free Ca^{2+} 93, 94–104, 254
    effect on phosphatidylinositol metabolism 130, 143
    effect on liver pyruvate dehydrogenase 161
    activation of glucose transport 165
    activation of glycogenolysis 231, 254
α_2-Adrenergic receptors
    inhibition of adenylate cyclase 56, 93, 125
    in platelets 125
β-Adrenergic receptors 91–94
    regulation of adenylate cyclase 44–45
    in cell fusion studies 48–49
    in liver 95, 103
    activation of glycogenolysis 231
Aequorin 86
Alanine
    as carbon source in liver 252, 264–265
    effect on pyruvate kinase 265
Alanine aminotransferase 252, 264

Aldolase 10, 250, 256, 257, 270
Allosteric regulation 19–20
Amiloride 199
Amino acids
    as carbon source in liver 210, 251–254, 264–265
Amino acid transport
    effect of insulin 145
    effect of growth factors 191
AMP
    regulation of phosphofructokinase 221–222, 224
    regulation of phosphorylase b 232
    level in fat cells 284
Amphotericin B 199
Amplification of signals 13, 22
Androgens 170
Angiotensin II
    activation of glucose release 91, 93
    effect on cytoplasmic free Ca^{2+} in liver 94–104
    effect on phosphatidylinositol metabolism 130
    effect on glycogenolysis in liver 254
Apolipoprotein C 213
Arabinose 226
Arachidonic acid 125
    as component of inositol phospholipids 131, 141
    activation of phospholipase C 134
Arginine 264
Arsenoazo III 86, 87
Aspartate
    equilibrium with oxaloacetate and malate 253
    as carbon source in liver 264
Aspartate aminotransferase 252, 253
    in gluconeogenesis 263–265
Antagonistic interactions between Ca^{2+} and cyclic AMP 124–127
ATP
    rate of turnover 136
    labelling with ^{32}P in intact cells 165
    inhibition of phosphofructokinase 221–222

inhibition of phosphorylase b 232
inhibition of citrate synthase 243, 253
inhibition of α-oxoglutarate dehydrogenase 243
inhibition of liver pyruvate kinase 265
limitation of fatty acid biosynthesis 276−279
ATP-citrate lyase (citrate cleavage enzyme) 250
in fatty acid biosynthesis 270−271, 273, 274, 278−280, 285, 286

Banting and Best 144
Blood glucose
   maintenance in starvation 210, 244
   effect on insulin release 226
   effect on glycogen synthetase in liver 254
   and liver glycolysis 255
   and liver gluconeogenesis 258
Blow fly salivary gland
   in phosphatidylinositol metabolism 127−128, 137
Brain
   carbon sources in starvation 214−215
Brown fat 268

Ca^{2+} ions
   as second messenger 33, 34, 83−84, 94−104
   activation of brain adenylate cyclase 57
   control of levels in resting cell 66−82
   concentration in cell compartments 66−67, 82, 84
   plasma membrane transport 67−70
   endoplasmic reticulum transport 70−71
   sarcoplasmic reticulum transport 71−73
   mitochondrial transport 73−80
   intramitochondrial free concentration 80
   measurement of free cytoplasmic levels 84−91
   activation of phosphorylase kinase 97, 101, 254
   binding proteins 105
   interaction with cyclic AMP response 117−129
   requirement for phosphatidylinositol metabolism 133−135
   as message in phosphatidylinositol metabolism 140−141
   as second messenger for insulin action 160−162
   in mitogenesis 200
   activation of PDH-phosphatase 239, 283−284
   activation of NAD-isocitrate dehydrogenase 243
   activation of α-oxoglutarate dehydrogenase 243
Ca^{2+}/ATPase
   in plasma membrane 68−69, 115, 163
   in endoplasmic reticulum 70
   in sarcoplasmic reticulum 71−73, 122
   role in control of cytoplasmic free Ca^{2+} 81, 82
Ca^{2+} buffering by mitochondria 78−80, 98−102
Ca^{2+} electrode 79, 84−85, 87
Ca^{2+} flux studies
   ^{45}Ca fluxes 89−91
   direct measurement 91
Ca^{2+} mobilizing hormones 83, 94−104, 118, 121
   effect on protein phosphorylation in liver 123
Ca^{2+} uptake
   across plasma membrane 67, 119
   across endoplasmic reticulum membrane 70, 119
   by sarcoplasmic reticulum 71, 119
   by mitochondria 73−76, 119
   significance for Ca^{2+} buffering 78−79
   role in hormone action 98−102
Ca^{2+} release
   across plasma membrane 68−70, 81, 91, 101, 119
   from endoplasmic reticulum 71
   from sarcoplasmic reticulum 73
   from mitochondria 76−79, 200
Ca^{2+} binding domain 107, 108
Calcium binding proteins 105−116
Calmodulin 106−116, 99, 117
   activation of cyclic AMP phosphodiesterase 38, 119, 121
   activation of brain adenylate cyclase 57, 114, 119
   inhibition of plasma membrane Ca^{2+} ATPase 67, 115, 121
   effect of dilution on response 105
   structure 106−108
   Ca^{2+} binding 108−111
   interaction with proteins 109, 112
   processes affected by 112−115
   protein kinases 115−116, 123−124, 165
   myosin light chain kinase 127
Carbohydrate absorption and transport 211−212
Carnitine shuttle 241, 245−246
Carrier proteins 1
Catalytic unit of adenylate cyclase 45
   interaction with guanine nucleotide binding proteins 52−53
   activation by calmodulin in brain 57, 114, 119
Catecholamines 29, 83
   receptors 91−94
   biosynthesis 191
Cathepsin B 196
CDP-diacylglycerol inositol transferase 132, 135
Cell cycle 192
Chloroquine 196, 205
Cholera toxin
   ADP-ribosylation of Gs 42−43
   lack of effect in cyc⁻ S49 lymphoma cells 46
   retrograde transport 202
Cholesterol 212
Cholesterol ester 212
Chromatin
   binding of steroid hormone receptors 180−181, 182−183
Chylomicrons 212−213, 287
Citrate 242

inhibition of
    phosphofructokinase
    222–223
  inhibition of citrate synthase
    243
Citrate cleavage enzyme
    (ATP-citrate lyase) 250
  in fatty acid biosynthesis
    270–271, 273, 274,
    278–280, 285, 286
Citrate synthase 242, 250, 251
  control of 243, 251–253
  in fatty acid biosynthesis
    270–271, 273, 274, 281
Citric acid cycle 242
  regulation in muscle 238–243
  in liver 245, 251–254
  in adipose tissue 276–281
Citrulline 264
Cl^- ions
  in salivary secretion 127–128
Clonidine 93
Coenzyme A (CoA)
  limitation of β-oxidation by
    240, 243
  in fatty acid biosynthesis
    269–272
Colchicine 153, 157, 196, 203
Collagen 124
Collision coupling hypothesis
    50–52
Concerted model of allosteric
    regulation 16, 19
Concanavalin A (Con A)
  effect on phosphatidylinositol
    metabolism 130
  as an insulin mimetic agent
    150, 162
  as a mitogen 200
Control strength 2, 4, 9
Chemical compartmentation 9
Chemiosmotic hypothesis 73–74
Cooperativity
  of substrate binding 16–19
  of insulin binding 147–148
Corticosterone 170
Creatine kinase 222
Creatine phosphate 222
Cross-over plot 8
Cyclic AMP 83, 165
  mechanism of action 59–65
  in glycogenolytic cascade
    23–24, 228–229
  as a second messenger 33–36

in control of liver metabolism
    94–95
effect on Na^+ dependent Ca^{2+}
    efflux from mitochondria 103
interaction with Ca^{2+} response
    117–129
effect on phosphatidylinositol
    kinase 135
as an insulin second messenger
    158–159
insulin/cyclic AMP dependent
    phosphorylation of liver
    plasma membrane 168
in control of mitogenesis 198
in nerve growth factor action
    205
Cyclic AMP phosphodiesterase 35,
    37–38, 117, 118
  effect of calmodulin 38, 113
  in blow fly salivary gland 128
  insulin/cyclic AMP dependent
    phosphorylation 168
Cyclic GMP 159
Cyclic GMP phosphodiesterase
    37–38, 113
Cysteine 149
Cytochalasin 153, 157, 206
  effect on glucose transport 16
Cytoplasmic free Ca^{2+} concentration
  measurement 84–91
  response to hormones in liver
    94–104
  effects of cyclic AMP 119–122
  in relation to
    phosphatidylinositol
    metabolism 130, 140–141
  control by
    inositol-1,4,5-trisphosphate
    104, 142
Cytoskeleton 115

Desensitization of hormone
    response
  of hormone sensitive adenylate
    cyclase 54–55
  of phosphatidylinositol
    response 138–139
  of insulin response 157
  of EGF response 196
  of NGF response 203
Deoxythymidine kinase 25
Diabetes 150
  ketosis in 249

Diacylglycerol 132, 141, 288
Diacylglycerol transferase 288
Diacylglycerol lipase 288–289
Dietary lipid
  absorption and metabolism
    212–214, 287
Dihydroalprenolol 93
5α-Dihydrotestosterone 174
Dihydroxyacetonephosphate 219
DNA
  binding of steroid receptors to
    180–181
  phosphorylation 198
  methylation 198
DNA replication
  activation by insulin 144–145,
    150, 157
  control by growth factors
    192–200
DNA transcription
  in control of protein synthesis
    126, 170
  activation by steroid receptors
    182, 183
Dopamine 93
Dopamine β-hydroxylase
  induction by NGF 191, 202,
    203
Dorsal root ganglia
  in studies of NGF action 186,
    191
  as source of NGF 188
  accumulation of NGF 202–203

EGTA
  effect on response of liver cells
    to hormones 95–96
Encephalins 57
Endoplasmic reticulum
  Ca^{2+} transport by 70–71
  as source of Ca^{2+} in response to
    hormones 98–104
  glucose transporters in
    163–165
Enolase 10, 256, 257, 270–271
Enzyme turnover 24–27
Epidermal growth factor (EGF)
    186–199
  effect on phosphatidylinositol
    metabolism 130
  amino acid sequence 186
  source 187–188
  structure 188–189

physiological effects 190–191
mechanism of action 192–200
EGF receptor 193–197
    structure 193
    internalization 193–194, 195–196
    tyrosine kinase activity 195, 196, 197
    effect of antibodies to 196
Epididymal fat pad 268
Equilibrium constant 2, 3
Erythrocytes
    turkey 48
    Ca^{2+}/ATPase in 68–69
    rat 119–121
Extrinsic control 4–6, 12, 13, 227

F^- ions
    activation of adenylate cyclase 57–58
    inhibition of protein phosphatase 231
Fat cells
    hormone receptors in 47
    insulin binding studies 146–147
    glucose transport 149
    insulin receptor antibodies and 150
    effects of insulin on 158–162, 165, 167
    acute hormone effects 268
Fatty acids
    as carbon source in muscle 210, 214, 219, 238–240
    as carbon source in liver 210–211, 244
    in absorption of dietary lipid 212
    in starvation 214–216, 244
    inhibition of glucose transport 227
Fatty acid biosynthesis
    in liver 249–251
    in adipose tissue 268–286
Fatty acid oxidation
    in muscle 240–241
    in liver 244–247
Fatty acid synthetase 250, 269–270, 272, 286
Fatty acylcarnitine 245–246

Fatty acyl CoA 241, 245–246, 254
    effect on acetyl CoA carboxylase 285
Fatty acyl CoA synthetase 241, 246
Feedback regulation 5–6, 19
Ferritin insulin 153
Fibrinogen 124
Fibroblasts
    as source of NGF 188
    in study of EGF function 192
Fibroblast derived growth factor (FDGF) 187, 193, 198, 199
Fibroblast growth factor (FGF) 187
Flux generating reaction 2, 3–4
Forskolin 58
Friend erythroleukemia cells 48
Fructokinase 212–213, 250
Fructose
    as carbon source in liver 210, 244, 249–251
    absorption and metabolism 211–213
    as carbon source in adipose tissue 269, 282–283
Fructose-1-phosphate 212
Fructose-6-phosphate 212–221
    activation of phosphofructokinase 223
    in PFK/fructose bisphosphatase substrate cycle 223–224
    equilibrium with G6P and G1P 224–225, 258
    as product of pentose phosphate pathway 276–277
Fructose-1,6-bisphosphate 221
    activation of phosphofructokinase 223
    in PFK/fructose bisphosphatase substrate cycle 223–224
    activation of liver pyruvate kinase 265
Fructose-1,6-bisphosphatase 256, 257
    in PFK/fructose bisphosphatase substrate cycle 20, 223–224
    in control of gluconeogenesis 259–261
Fructose-2,6-bisphosphate 259–261, 284
Fructose-2,6-bisphosphatase
    regulation by cyclic AMP dependent protein kinase 60, 124

    in regulation of gluconeogenesis 259–261
Fructose-6-phosphate/ fructose-1,6-bisphosphate substrate cycle 20
    in muscle 223–224
    in liver 259–260, 267
Fumarase 242, 252, 264, 281
Fumarate 242, 252, 264

Galactose 211–212
β-Galactosidase 149
GDP
    inhibition of adenylate cyclase 44–45
GDPβS 40
Glioma cells 188
Glucocorticoids 170
    receptor 183
    in starvation 244, 259, 266
Glucagon
    activation of adenylate cyclase 39–47
    effects in liver 91, 94, 103, 121–124
    antagonism of insulin effect 158–159
    in starvation 244, 259, 265, 266
    regulation on ketone body synthesis 247
    regulation of liver glycogenolysis 254
Glucokinase 14, 25, 250, 256–259
Gluconeogenic cascade 259–261
Gluconeogenic shuttles 261–265
Gluconeogenesis 26, 91, 170, 210, 216, 244, 247
    citrate cycle as a carbon source 251–253
    pathway and regulation 255–267
Glucose
    release from liver 94–104, 210
    as carbon source for muscle 210, 219
    absorption in intestine 211–213
    as carbon source in liver 244
    as carbon source in adipose tissue 269
    as substrate for pentose phosphate pathway 275

Glucose-1-phosphate 212–213, 224, 259
Glucose-1-phosphate uridyltransferase 228
Glucose-6-phosphate 1, 219, 227, 256
    inhibition of muscle hexokinase 224
    control of glucose uptake 224–226
    in regulation of glycogen metabolism 228, 235, 254
    fate in liver 255
    in regulation of gluconeogenesis 259
    in pentose phosphate pathway 275
Glucose-6-phosphatase 227, 256–259
Glucose-6-phosphate dehydrogenase 275, 286
Glucose-6-phosphate isomerase 10, 224, 250, 256, 257, 270
Glucose-6-phosphate/glucose substrate cycle 258–259, 267
Glucose transport 7, 12, 221, 149
    regulation by insulin 144–145, 162–165
    activation by growth factors 191
    in intestinal lumen 212
    control in muscle 226–227
    in liver 256–258
    in fatty acid biosynthesis 270, 282–283, 284
Glutamate 252, 263, 264
Glutamate dehydrogenase 252
Glyceraldehyde-3-phosphate 219, 276, 277
Glyceraldehyde-3-phosphate dehydrogenase 10, 219, 256, 257, 262, 270, 273
Glycerol 136, 215, 217, 287
Glycerol kinase 287
Glycerol-3-phosphate 219, 287–288
Glycerol phosphate acyltransferase 287–288
Glycerol phosphate dehydrogenase 270–271
Glycogen 210, 219
    in muscle 227–237
    in liver 244, 254

in adipose tissue 282–283
Glycogen synthetase 228–284
    phosphorylation in liver 123, 254
    effect of insulin on phosphorylation 145, 166–167
    effect of peptide regulator on phosphorylation 162
    regulation in muscle 233–235
Glycogen synthetase kinase 3 62, 233
    and insulin action 167, 235
Glycogen synthetase kinase 4 233–234
Glycogen synthetase kinase 5 233–234
Glycogenolysis 35, 91, 105, 214, 216
    in muscle 227–233
Glycogen synthesis
    in muscle 226–227, 233–237
    in liver 244
Glycogenolytic cascade 23–24, 228–229
Glycolysis 212
    in heart muscle 10–12, 219–227
    in insect flight muscle 12
    in white muscle 214
    G6P as a signal of flux rate 235
    as a source of acetyl CoA 238
    in liver 255–267
    in adipose tissue 278–280, 284
Gramicidin 199
Growth factors 186–208
Growth factor receptors
    Internalization 33
    EGF 193–197
    NGF 201–207
GTP
    regulation of adenylate cyclase 38–45
    effect of glucagon binding 38
    contamination of ATP 39
    adenylate cyclase activation cycle 41–42
    requirement for inhibition of adenylate cyclase 56
GTPγS 40, 41
Guanine nucleotide binding proteins 45, 46, 52–53, 57, 125

Guanylyl imidodiphosphate(p(NH)ppG) 40–45, 50–52
Guanylyl methylenediphosphate 40, 41

Heart muscle 214–215
    glycolysis in 10–12, 219–227
    Effects of β-adrenergic agonists 122
    Use of dietary lipid 213
    Energy metabolism 218
Hexokinase 10, 19, 212, 219, 221, 224–226, 257, 270, 286
Histamine 130
Histones 181
Hormone binding 31–33, 145–148
Hormone receptors 29
    relationship of occupancy to physiological response 31–33, 139
    as component of adenylate cyclase 45, 47–53
β-Hydroxybutyrate 241, 247–249
6-Hydroxydopamine 203
β-Hydroxymethyl glutaryl CoA (HMG CoA) 247–249
HMG-CoA lyase 247–249
HMG-CoA synthase 247–249
HMG-CoA reductase 25
5-Hydroxytryptamine (5HT)
    in platelet release reaction 124–125
    in blow fly salivary gland 127–128
    and phosphatidylinositol metabolism 130, 137–138

Information transfer 29
Inhibitor 1 60, 236
Inositol 132, 136, 137
Inositol-1,4-bisphosphate 132, 137
Inositol-1-phosphate 132, 137
Inositol-1,4,5-trisphosphate
    as a second messenger in liver 104, 142
    formation 132, 137–138
Insect flight muscle 12, 224
Insulin 227, 259, 268
    peptide factor 33, 162
    and Ca^{2+} ATPase 68, 162

effects on cell metabolism
144–145
mechanism of action 145–169
ferritin insulin 153
growth control 154–155, 193,
196, 198
second messenger 155–162
glucose transport and
144–145, 162–165,
226, 282, 284
protein phosphorylation
165–168, 239, 249,
283–284, 284–285
control of glycogen synthesis
228, 254
effect on ketone body
production 239
regulation of fatty acid
biosynthesis 278, 282–286
effect on lipoprotein lipase 287
effect on triacylglycerol
esterification 287–288
inhibition of lipolysis 144–145,
289
Insulin receptor
insulin binding 145–147
structure 148–150
cross-linking 150–153
antibodies 152–153, 245
tyrosine kinase activity 154,
168, 196
Insulin antibodies 152–153
Insulin-like growth factors (IGF)
146, 154–155, 187
Insulin receptor antibody
insulin mimetic effect
150–153, 162, 245
Insulin regulatory peptide 33, 162
Intestinal lumen
monosaccharide carriers in
150–153, 162, 245
Intrinsic control 4–6, 12, 13
Irreversible reactions 7
Isobutylmethylxanthine (IBMX)
198
Isocitrate 242
NAD-isocitrate dehydrogenase 242,
243, 252
NADP-isocitrate dehydrogenase 281
Isoprenaline 92, 93

K^+ ions
cytoplasmic and extracellular
concentration 67

in sarcoplasmic reticulum
Ca^{2+}/ATPase function 72
use with valinomycin to
generate a membrane
potential 76
in salivary secretion 127
K_m 13, 14–16
K_s 14
Kacser and Burns 2, 4
Ketone bodies
as a carbon source for muscle
210, 219, 240
production 210, 211, 244,
247–249
use by brain 214
physiological importance
248–249

Lactate
as a carbon source in liver 210,
244, 249–251
as a carbon source for
gluconeogenesis 262–264
release from adipose tissue
282–283
Lactate dehydrogenase 250,
262–264
Leiotonin 106
Lipid oxidation
in muscle 239–241
in liver 244–247
Lipid uptake in adipose tissue
286–287
Lipolysis
inhibition by insulin 144–145,
158–159
activation in starvation 245
regulation 288–289
Lipoproteins
production in liver 210
as a carbon source in muscle
240–241
as a source of fatty acids in
adipose tissue 268, 287
Lipoprotein lipase 213–214
in muscle 240–241
in adipose tissue 287
Lithium 137, 160
Liver
physiological function
210–211
processing of dietary lipid
212–214

regulation of metabolism
244–246
Liver cells
effects of hormones on
cytoplasmic Ca^{2+} 94–104
protein phosphorylation 118,
165–168
effect of cyclic AMP on Ca^{2+}
123–124
phosphatidylinositol
metabolism 139
effect of insulin on cyclic AMP
levels 158–159
effect of insulin on cytoplasmic
free Ca^{2+} 160–161
Lymphocytes 87, 152, 200
S49-lymphoma cells 46
Lysosomes 26, 196, 205, 206

Malate 242
as a carbon source for
gluconeogenesis 251, 252,
254
in fatty acid biosynthesis
271–274, 278, 280
NAD-malate dehydrogenase 242,
250, 252, 256, 257, 263, 264
in fatty acid biosynthesis
270-271, 273, 274, 278–280
NADP-malate dehydrogenase 250
in fatty acid biosynthesis 270,
271, 273, 281, 286
Malonyl CoA 245, 268–270
Mass action ratio 2–3, 7, 8, 9
Mellitin 199
Metabolic integration 209–217
Methoxamine 93
Methylamine 196
Methylation of proteins 21
3-O-Methylglucose 163, 206
Mg^{2+} ions
effect on inhibition of
adenylate cyclase 56
cytoplasmic and extracellular
concentration 67
role in sarcoplasmic reticulum
Ca^{2+}/ATPase function 72
effect on mitochondrial Ca^{2+}
buffering 79
effect on calmodulin 108
Michaelis and Menten 13
Microtubules 115, 153, 157, 196,
203
Mineralocorticoids 170

Mitchell 73
Mitochondria
    Ca^{2+} uptake 73–76
    Ca^{2+} release 76–78
    free Ca^{2+} concentration 80
    Ca^{2+} buffering 78–79
    as source of Ca^{2+} in response to hormones 98–104
Mitogenic response 195–200
Mitotic spindle 115
Mn^{2+} ions
    activation of adenylate cyclase 46
    effect on calmodulin 108
Mobile receptor hypothesis 47–49
Monensin 199
Monoacylglycerol 212
Monoacylglycerol lipase 288–289
Monoacylglycerol phosphate acyl transferase 288
Monoclonal antibodies 179
Multisite phosphorylation 233–236
Muscle 209–210, 213, 214–215
Muscle contraction 1, 105, 160, 165
Myosin light chain kinase
    regulation by cyclic AMP dependent protein kinase 60, 127
    regulation by troponin C 230

NADH
    inhibition of PDH 238
    in regulation of liver citrate cycle 253
NADPH
    Production for fatty acid biosynthesis 271, 273–276, 285–286
NECA (5' N-ethylcarboxyamide-adenosine) 198
MEN (N-ethyl maleimide) 48, 149
Nerve growth factor (NGF)
    and phosphatidylinositol metabolism 130
    discovery 186
    source 187–188
    structure 188–190
    physiological effects 191
    mechanism of action 201–208
NGF receptor 201–207
Neuroblastoma cells 188
Na^+ ions
    as second messenger 33

    in inhibition of adenylate cyclase 56
    cytoplasmic and extracellular concentration 67
    effect on endoplasmic reticulum Ca^{2+} uptake 70
    effect on Ca^{2+} efflux from mitochondria 77
    in regulation of mitogenesis 199–200
    in intestinal monosaccharide transport 211–212
Na^+/Ca^{2+} exchanger 69–70, 81, 125–126
    role in control of cytoplasmic free Ca^{2+} 81
    Ca^{2+}/Ca^{2+} exchange 90–91
    in mitogenesis 200
Na^+/H^+ exchange
    in mitochondria 77
    in plasma membrane 199–20
Na^+/K^+ ATPase
    activation by cyclic AMP in smooth muscle 125–126
    activation by growth factors 190–199
    as a source of energy for transport 211–212
Negative cooperativity 16, 147–148
Neuraminidase 149
Neurotransmitters 57
Neutral proteases 26
Newsholme 2
Noradrenaline 92
NTA 78–79
Nuclear acceptor site 181–183
Nuclear matrix 183
Null point titration 87–89, 96

Obelin 86
Occupancy theory of hormone action 31
Oestrogens 170
Oestrogen receptor 179
Oligomycin 161
Opiates 130
Ornithine 264
Ornithine decarboxylase
    rate of turnover in liver 25
    induction by EGF 190
Ouabain 199
Oxaloacetate 9, 242, 243, 251, 254
    regulation of liver citrate synthase 252–253

    as a carbon source for gluconeogenesis 262–265
    in fatty acid biosynthesis 271–273, 278, 280
β-Oxidation 238, 241
    regulation in muscle 239–240
    regulation in liver 245–247
α-Oxoglutarate 242, 251–253, 263–264
α-Oxoglutarate dehydrogenase 242, 243, 252, 282

Palmitate 269, 272
Palmitoyl CoA 269
Parvalbumen 106
PC12 phaeochromocytoma cells 201, 202, 206
PDGF receptor 196
Pentose phosphate pathway 274–277
    regulation 285–286
Peroxide 149
pH electrode 85
Phenoxybenzamine 93
Phenylalanine hydroxylase 60
Phenylephrine 92, 93
Phorbol esters 141, 193, 199
Phosphate
    role in mitochondrial Ca^{2+} uptake 75
    effect on intramitochondrial free Ca^{2+} 77, 80
    effect on mitochondrial Ca^{2+} buffering point 79
    activation of phosphofructokinase 222
    activation of phosphorylase b 233
Phosphatidic acid 132, 288
Phosphatidic acid phosphatase 288
Phosphatidylinositol
    cellular location 130
    structure 131
    synthesis and breakdown cycle 132
    effects of hormones on metabolism 135–142
Phosphatidylinositol-4-phosphate 137–138
    structure 131
    synthesis 132
    breakdown 134
Phosphatidylinositol-4,5-bisphosphate 104

structure 131
synthesis 132
breakdown 134–137
Ca^{2+} binding 141
Phosphatidylinositol/inositol
　exchange 132
Phosphatidylinositol kinase 132,
　134–135
Phosphatidylinositol
　phosphodiesterase 132,
　133–134
Phosphoenolpyruvate 9, 261–265
Phosphoenolpyruvate
　carboxykinase (PEP-CK)
　rate of turnover in liver 25
　effect of insulin 27
　in gluconeogenesis 253, 254,
　　256, 257, 261–266
Phosphofructokinase 250, 256, 257,
　270, 284
　in F6P/F-1, 6-bisP substrate
　　cycle 20, 212
　in control of muscle glycolysis
　　220–224
　in control of gluconeogenesis
　　259–261
Phosphofructokinase 2
　regulation by cyclic AMP
　　dependent protein kinase 60
　phosphorylation in liver 124
　in regulation of
　　gluconeogenesis 259–261
Phosphoglucomutase 224
6-Phosphogluconate dehydrogenase
　275, 286
6-Phosphogluconolactone 275
Phosphoglucoisomerase 10
Phosphoglycerate kinase 10, 219,
　256, 257, 270–271
Phosphoglycerate mutase 10,
　270–271
Phospholamban 122
Phospholipase C
　(phosphatidylinositol
　　phosphodiesterase) 132,
　　133–134
Phosphomonoesterases 132, 134,
　137
4-Phosphopantetheine 269
Phosphorylase kinase 229–233
　in glycogenolysis 23
　regulation by cyclic AMP
　　dependent protein kinase 60,
　　229–230
　and calmodulin 107, 229–230

and troponin C 116, 230
in liver 123–124
and insulin 161
phosphorylation in vivo 230
proteolysis of 232
phosphorylation of glycogen
　synthetase 233, 234, 254
Phosphorylase
　in muscle 22–23, 228,
　　232–233
　in liver 124
Platelet activating factor 130
Platelet release reaction 124–125
Platelet derived growth factor
　(PDGF) 187, 199
Platelet factor 4 124
Polyamines 190
Polypeptide hormones 29
Positive feed-forward regulation 5
Prazosin 92–93
Promoter region of genes 183
Progesterone receptor 179–182
Propranolol 92–93
Prostacyclin 125
Prostaglandins
　regulation of adenylate cyclase
　　55, 57, 93, 125
　and phosphatidylinositol
　　metabolism 141
Protein degradation 26–27, 216
Protein kinases 21–24
　calmodulin regulated 115–116,
　　123–124
　and insulin action 168
Protein kinase C 141, 143, 165
Protein kinase (cyclic AMP
　dependent) 22, 59–65, 115,
　117, 118, 124, 125–127,
　165, 233–234, 236, 238,
　244, 254, 265, 285, 288, 289
Protein kinase (cyclic GMP
　dependent) 62
protein kinase inhibitor 64
Protein phosphatases 21, 24, 167,
　234, 236
Protein phosphorylation 21–24
Protein synthesis
　in enzyme turnover 26
　effect of insulin on 145
　effect of starvation on 170
　activation by NGF 201
Proteolysis
　regulation of enzyme activity
　　by 21, 232
Purine and pyrimidine transport 191

Pyruvate 9, 21
　inhibition of PDH kinase 238
　as carbon source for fatty acid
　　biosynthesis 269, 282–283
Pyruvate/malate cycle 271–274,
　278, 280
Pyruvate/oxaloacetate/
　phosphoenolpyruvate
　　substrate cycle 265–267
Pyruvate carboxylase 250, 252, 256,
　257
　in gluconeogenesis 261–265,
　　265–266
　in fatty acid biosynthesis
　　270–271, 273, 278–280,
　　281, 282
Pyruvate decarboxylase 238–239
Pyruvate dehydrogenase (PDH) 250
　effect of insulin on
　　phosphorylation 145, 161,
　　166
　activation by insulin regulatory
　　factor 162
　regulation in muscle 238–239
　in fatty acid biosynthesis
　　269–274, 278–280, 281,
　　283–284
Pyruvate kinase 9, 10, 250,
　270–271
　regulation in liver 60, 256, 257,
　　260, 261, 265
PDH-kinase 238–239
PDH-phosphatase 238–239

Quin 2 86–87
　measurement of cytoplasmic
　　free Ca^{2+} in liver 96, 99, 121
　measurement of cytoplasmic
　　free Ca^{2+} in lymphocytes 200

Rabin model 18–19
Rate controlling step
　identification 6–12
　in relation to V_{max} and K_m 16
Rate-limiting step 1
Rate theory of hormone action 31
Rb^+ ions 199
Remnant particle 214
Retrograde transport 202–203, 204
Ribose-5-phosphate 276, 277
Ribulose-5-phosphate 275–276,
　277
RNA-polymerase 25

mRNA 182–183
mRNA translation 26, 144–145, 170
Rous sarcoma virus 196–197
Ruthenium red 76

Sarcoma (mouse)
    as source of NGF 186
Sarcoplasmic reticulum
    Ca^{2+} transport by 71–73
    phosphorylation in smooth muscle 126–127
Scatchard analysis 148
Schramm 48
Second messenger 30–33
    criteria for 30, 156
    cyclic AMP as 36–37
    Ca^{2+} as 94–104
    inositol-1,4,5,-trisphosphate as 142
    for insulin 155–162
    as products of phosphatidylinositol breakdown 139–142
Second messenger hypothesis 30
    related to cyclic AMP 35–36
    related to Ca^{2+} ions 84
Sendai virus 48
Sequential model of allosteric regulation 17–19
Serine 22
Serine dehydratase 25
Serum
    Requirement for mammalian cell culture 186
smooth muscle
    interactions between Ca^{2+} and cyclic AMP 125–127
    glycolysis in 214
    glycogen in 227
Somatomedins 187
Squid giant axon 69
Starvation 210, 214–217
Static model of adenylate cyclase regulation 50–51
Stearic acid 131, 269
Steroid hormones 29, 170–185
    structure 170–171
    entry into cell 172–173
    acute effects 173
    modification 174
    receptors 174–183
Steroid hormone receptors
    cell surface 173

hormone binding 174–176
meroreceptors 175, 185
structure and function 175–182
progesterone 179–180
cellular location 174–179
nuclear binding 180
reversal of response 183–185
Submaxilliary gland of male mouse
    as source of EGF and NGF 186–188
Substrate control 13–16, 257–258
Substrate cycles 20–21
    fructose-6-phosphate/fructose-1,6-bisphosphate 223–224, 259–260
    glucose/glucose-6-phosphate 258–259
    pyruvate/oxaloacetate/phosphoenolpyruvate 269
Succinate 241, 242, 252
Succinate thiokinase 242, 252, 281
Succinate dehydrogenase 242, 252
Succinyl CoA 241, 242, 243, 252
    in regulation of ketone body synthesis 247
Succinyl CoA-3-oxoacid CoA transferase 241
Sutherland 30, 36

3T3 cells
    use in study of mitogenesis 192, 193, 197, 198
T-cell growth factor IL-2 187
Testosterone 174
Testosterone-5-reductase 174
Tetanus toxin 202
Tetradecanoylphorbol-13-acetate (TPA) 193, 194, 200
Threonine 22
Threonine deaminase 5
Thrombin 125, 130
Thyroid hormones 30
Tolkovsky and Levitzki 50
Transaldolase 276–277
Transketolase 276–277
Triacylglycerol
    as a component of chylomicrons 212
    synthesis in liver 216, 244
    synthesis in adipose tissue 216, 268, 287–288
    storage in muscle 219, 240
Triacylglycerol lipase

regulation by cyclic AMP dependent protein kinase 60, 288, 289
Triose kinase 213, 250
Triose phosphate isomerase 10, 213, 250, 256, 257, 270
Troponin I 60
Troponin C 106, 116, 123
Tryptophan oxygenase 25
Turkey erythrocyte 48
Turkey erythrocyte plasma membrane 43, 44–45
Tyrosine kinases 22
    insulin receptor 154, 168
    EGF receptor 195–197
Tyrosine aminotransferase 25
Tyrosine hydroxylase
    induction by NGF 191, 201, 203

Uncouplers 161, 165
Urea 210
Urea cycle 264–265
Urogastrone 188

V_{max} 14–16
Valinomycin 76
Vasopressin 91, 93
    effect on cytoplasmic free Ca^{2+} in liver 94–104
    effect on protein phosphorylation in liver 116, 123, 124
    effect on phophatidylinositol metabolism 130
    effect on liver PDH 161
    effect on glucose transport 165
    in control of cell growth 193, 198–200
Verapamil 67
Vinblastine 153, 203
Viral genes
    in regulation of cell growth 196–197
VLDL (very low density lipoproteins) 214
Voltage dependent Ca^{2+} channel 67, 122

Xylulose-5-phosphate 276, 277

Zn^{2+} ions 180–190